Houghton Mifflin Action Series
A Reading Breakthrough Program

Encounters

Challenges

Forces

Crosscurrents

Ventures

Vibrations

Reactions

Counterpoints

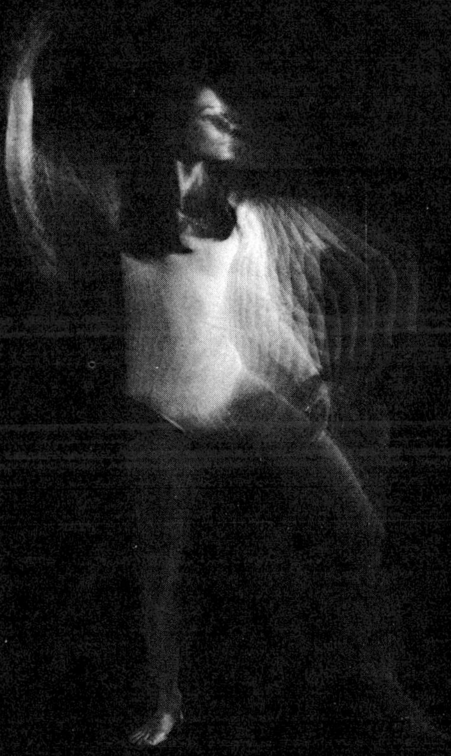

Houghton Mifflin Action Series

Vibrations

Harvey R. Granite *Rochester Public Schools*

Millard H. Black *Los Angeles City School District*

Jo M. Stanchfield *Occidental College*

Houghton Mifflin Company • Boston
New York • Atlanta • Geneva, Illinois • Dallas • Palo Alto

The editors and publisher wish to express their appreciation to Mrs. Marjorie Farmer for her helpful advice during the preparation of this volume.

Library of Congress Catalog Card Number: 79-147101

Hardcover Edition ISBN: 0–395–12052–7
Softcover Edition ISBN: 0–395–12053–5

Acknowledgments

"Adventures of Isabel," copyright 1936 by Ogden Nash. From *Family Reunion* by Ogden Nash, by permission of Little, Brown and Co.

"All the Years of Her Life," copyright 1935, 1962 by Morley Callaghan. Originally appeared in *The New Yorker.* Reprinted by permission of the Harold Matson Company, Inc.

"Ashes for the Wind," from *Spanish Stories and Tales,* edited by Harriet de Onis. Copyright 1954 by Alfred A. Knopf, Inc. Reprinted by permission of the publisher.

"Bessie Black Bull," from the book *When the Legends Die* by Hal Borland. Copyright © 1963 by Hal Borland. Reprinted by permission of J.B. Lippincott Company.

"Friendship," from *The Night Visitor and Other Stories* by B. Traven. Copyright © 1966 by B. Traven. Reprinted by permission of Hill and Wang, Inc.

"How the Snake Got His Rattles," by Julius Lester, from *Black Folktales.* Copyright © 1969 by Julius Lester. Reprinted by permission of the Richard W. Baron Publishing Co., Inc.

"Indian Sky," from the book *The Selected Poems of Alfred Kreymborg 1912–1944* by Alfred Kreymborg. Copyright 1945 by Alfred Kreymborg. Published by E.P. Dutton & Co., Inc., and reprinted with their permission.

"Jug of Silver," copyright 1945 by Truman Capote. Reprinted from *A Tree of Night and Other Stories* by Truman Capote, by permission of Random House, Inc.

"The Little Girl and the Wolf," copyright 1940 James Thurber. Copyright 1968 Helen Thurber. From *Fables for Our Time,* published by Harper and Row. Originally printed in *The New Yorker.*

"Lost Sister," copyright © 1956 by Dorothy Johnson. Appeared originally in *Collier's.* Reprinted by permission of McIntosh and Otis, Inc.

"My Elder Brother," from *Childhood, Boyhood, Youth* by Leo Tolstoy, translated by Rosemary Edmonds. Reprinted by permission of Penguin Books Ltd.

"Not to Go with the Others," from *Here To Stay* by John Hersey. Copyright © 1962 by John Hersey. Reprinted by permission of Alfred A. Knopf, Inc.

"The Rabbits Who Caused All the Trouble," copyright 1940 James Thurber. Copyright © 1968 Helen Thurber. From *Fables for Our Time,* published by Harper & Row. Originally printed in *The New Yorker.*

"The Talking Skull," a Nupe folk tale. From *African Genesis* by Leo Frobenius and Douglas G. Fox. Published by Stackpole Books. Copyright 1937.

"The Tiny Closet," reprinted by permission of International Famous Agency. Copyright © 1950, 1953, 1954, 1955, 1958, 1960, 1962, by William Inge.

"Tom Edison's Shaggy Dog," reprinted from *Welcome to the Monkey House* by Kurt Vonnegut, Jr. Copyright © 1950, 1951, 1953, 1954, 1955, 1956, 1958, 1960, 1961, 1964, 1966, 1968 by Kurt Vonnegut, Jr. A Seymour Lawrence Book/Delacorte Press. Used by permission.

"The Use of Force," from *The Farmers' Daughters* by William Carlos Williams. Copyright 1938 by William Carlos Williams. Reprinted by permission of New Directions Publishing Corporation. (Punctuation altered for this book.)

"Where," from *Doubleheader—Hurrah for Anything* by Kenneth Patchen. Copyright © 1958 by Kenneth Patchen. Reprinted by permission of New Directions Publishing Corporation.

"The Wishing Well," by Phillip Bonosky © 1949 Masses & Mainstream, Inc.

"Your Poem, Man," by Edward Lueders. From *Some Haystacks Don't Even Have Any Needle* by Stephen Dunning, Edward Lueders and Hugh Smith. Copyright © 1969 by Scott, Foresman and Company.

Contents

In anger or fear, we often destroy what is most important to us.

Friendship

B. TRAVEN

One early afternoon the French owner of a restaurant located on Bolivar Street in Mexico City noticed a medium-sized black dog sitting on the sidewalk near the open door of his establishment. Looking at the Frenchman with soft brown eyes which sparkled with a suggestion of making friends, the dog put on that good-natured, innocent, somewhat tragicomic expression found on some old tramps who, no matter what happens to them, always have a humorous answer in store, even if they are kicked down the porch steps and a bucket of water is thrown in their face.

For a moment the Frenchman glanced at the dog. Interpreting that look in his own way, the dog wagged his tail, cocked his head, and gave his open mouth such a quizzical twist that he appeared to be grinning at the Frenchman.

So he couldn't help but smile back, and for an instant he had the strange sensation that a little spark of the golden sun had crept into his heart as if to touch it and warm it.

His tail wagging faster now, the dog rose slightly, sat down again, and remaining in this sitting position, moved a few inches closer to the door, yet did not enter.

Considering this a very decent attitude for a hungry street dog, the Frenchman could resist no longer. From a half-empty plate, just

taken from a table by a waitress walking past him toward the kitchen, he picked up a steak which the customer, obviously not hungry, had only nibbled.

Lifting that juicy steak with two fingers, holding it high for a few seconds, his eyes fixed on the dog, he waved it invitingly, and by a move of his head, hinted at the dog to come inside and get it. The dog, wagging not only his tail but his whole hindquarters, closed and opened his mouth quickly, licking his lips with a rosy tongue. Still, he would not enter.

The Frenchman, becoming more interested in the dog than in his patrons, left his place behind the counter and took the steak to the door, playing it before the dog's nose for a few moments before he let it go to its final destination.

The dog caught it more gently than hastily, gave the donor a grateful look, stepped back from the door, and lay down on the sidewalk close to the café. There he ate that big steak in the undisturbed manner that comes only from a clear conscience.

Done with his meal, the dog stood up, approached the café door, sat down, and waited patiently until the Frenchman would once again notice him. As soon as the Frenchman granted him that hoped-for look, the dog rose, wagged his tail, put on his comical grin, shook his head as if in fun so that his ears fairly shook, turned, and went on his way.

The Frenchman, naturally, thought that the dog had returned to face him hoping for another helping. But when he came to the door, holding between his fingers a chicken leg with lots of meat on it, the dog was gone. So he concluded that the dog had appeared a second time for no other reason than to say "Thanks a lot, *monsieur*"; otherwise, he would have waited until given another mouthful.

The Frenchman quickly forgot the incident, believing the animal to be only another street dog that visited café doors, searching for food, begging diners for a roll or a bone, until driven away by the waitresses.

Next day, however, about the same time, at half past three, the dog was sitting again at the same place by the open door. The Frenchman, seeing him seated there, smiled at him as if he were an old acquaintance. The dog smiled back, with a very funny expression of silent laughter. When he caught the Frenchman's friendly

smile of welcome, he got halfway up, just as he had done the day before, wagged his tail in greeting, and widened his grin with his pink tongue dangling over his lower jaw.

With a backward jerk of his head, the Frenchman invited the dog to come inside and have his free lunch close to the counter. The dog, however, who clearly had grasped the meaning of that gesture, came only one short step nearer the door — but enter he would not. The Frenchman realized that the dog refrained from coming inside not so much out of fear as from the apparently inborn decency of intelligent dogs who sense that a room in which humans dwell is not the proper place for dogs living on and by the street.

The Frenchman raised his hand to the countertop, drummed his fingers, and looked at the dog, trying to make him understand that he should wait a few minutes for a plate with a tasty bone on it to come back from one of the tables. To the man's great surprise, the dog understood this finger language perfectly. He stepped back a little from the door, as if not to molest patrons coming in or going out. He lay down on the sidewalk, his head between his forepaws, and with eyes half closed, he watched the Frenchman attending customers at the counter and collecting money brought by the waitresses to the cash register.

Five minutes later, when a waitress carried a tray of dishes from the tables, the owner winked at her, picked up a huge leftover sirloin T-bone, took it out to the dog, played it before his nose, and let go.

The dog took it from the man's hand as gently as if it had been offered by a child. And as he had done the day before, he lay down on the sidewalk close to the café and there enjoyed his lunch without hurry.

Now, remembering the dog's peculiar attitude of yesterday, the Frenchman was anxious to learn what the dog would do today, once he had finished his bone. He wished to know whether yesterday the dog had acted in that particular way only because of an occasional impulse or because of natural good manners or by some sort of training.

He was just about to bet a patron that the dog would come to the door to say "Thank you" when he noticed a shadow on the floor near the entrance. He watched the dog from one corner of his eye, purposely avoiding a full look at the animal now sitting by the door and patiently waiting to be noticed by the owner. But the Frenchman busied himself behind the counter and at the cash register, all the while watching the dog by stealth just to see how long he would remain sitting there hoping to get his chance to say "Thank you very much, until tomorrow."

Three or four minutes had passed before the Frenchman decided to look up and gaze straight at the dog's face. Right away, and as if the dog had been ordered to do just that, he rose, wagged his tail, put on his quizzical grin, turned around, and disappeared.

From that day on, the Frenchman always had a juicy piece of leftover meat ready for the dog. And he came every day, appearing at the door so punctually that one could set one's watch by his arrival. Always at half past three the Frenchman would direct a glance towards the door and see the dog sitting there greeting the owner with his drollish grin.

And so it went for six or seven weeks without the slightest change in pattern. In fact, the Frenchman came to regard the black street dog as his most faithful customer, and he considered him a sort of mascot.

In spite of the fact that the dog by now felt positively assured of the Frenchman's sincere friendship, he did not relax his formal manners. Never once had he come inside, no matter how often or how insistently the friendly owner invited him to enter.

The Frenchman grew very fond of the dog. He wanted him to hang around for good, make himself useful by chasing less decent dogs away, and guard the premises night and day — or just wait around and have a good time. He, of course, didn't know whether the dog had a master or lived on his own. Lately he had begun to caress the dog for a few moments when he gave him his lunch. It seemed the dog had never in his life known what love was, so he liked that sign of true affection. Yes, he relished the caresses. While he was stroked and patted, he would wait patiently, a steak in his mouth, until the Frenchman ceased petting him and returned to his counter. Only then, and not before, would the dog step back from the door, lie down in his accustomed place on the sidewalk, and eat his meal. When finished, he would come back to the door to wag his tail, grin at the Frenchman, thereby telling him in his own way, "*Merci beaucoup,*[1] *monsieur,* same time tomorrow, bye-bye," and then turn and trot away.

Came a certain day when the Frenchman had a terrific argument with a patron who had been served a hard roll and who, setting his teeth into that roll for a hearty bite, had broken a tooth.

The Frenchman was angry with the waitress who, poor girl, was fired immediately and now crouched in a dark corner weeping bitterly. It hadn't been her fault, entirely. Of course she ought to have noticed that the roll was hard like wood, but so should the customer have noticed it. Anyway, it wouldn't be hygienic or mannerly for a waitress to take each roll into her hand and squeeze it to test its freshness before serving it. But she had served that roll, and so she was blamed for what happened.

The real culprit, though, was the baker who had intentionally or carelessly thrown that hard old roll among the hot fresh ones. Anyhow, the damage was done.

The Frenchman, in a thundering rage, rang up the baker and told him that he was on his way, loaded gun in hand, to kill that

[1]*Merci beaucoup* (mär·sē′ bō̄·kü′): French for "Thank you very much."

careless dough-kneader like the rat that he was, for he never was anything else but a stinking sewer rat whose relationship to his father, to say the least, was questionable at all accounts; whereupon the baker told the Frenchman a few of those little mind-openers which, if uttered inside a church, would make its white-washed walls turn deep red and stay red for keeps. This lively exchange of personal opinions ended with the Frenchman hanging up the receiver with such force that — had telephone engineers not foreseen attacks of this kind and calculated almost perfectly the force which a maddened telephone-user might employ — nothing of the apparatus would have been left intact. As it was, only the hook was bent and the wall plaster nicked.

His face red as a ripe tomato, two bluish veins protruding from his hot forehead, the Frenchman returned to the counter, and looking up, saw his friend the black street dog sitting by the door, grinning, amused as ever, waiting for his lunch to be served.

Seeing that dog sitting there so quiet and innocent, apparently unbothered by any of the worries which make café owners appear to be twenty years older than their true age, gazing at the animal as if spying it for the first time in his life, seeing him eagerly wag his tail to greet his benefactor and put on his comical grin to please the friend who liked that doggy face so very much, the Frenchman — practically blinded with madness and driven by a sudden brutal impulse which he, who had no violent temper, later could never explain to himself — took that hard roll from the counter and threw it with all his force at the dog.

The dog had clearly observed the Frenchman's move. He had watched the angry man pick up the roll and had no doubt about the roll's being aimed at him. He had seen it shooting through the air towards him. He could easily have dodged that roll had he wished to, for he was a dog that lived entirely on what the street offered him, and he was used to the hard life without a master or with a master so poor that he could give little else to his dog but love. Surely the dog had cleverly dodged, perhaps hundreds of times, the sticks and stones thrown at him, and he certainly was an expert at avoiding any sort of missile.

A slight movement of the head would have sufficed[2] to save him from being hit by that hard roll. Yet he did not move. Not an inch did he move. He kept his soft warm eyes fixed straight at the Frenchman's face without a flicker. And without a tremor he most bravely took the shot.

For several seconds he kept sitting in the same place as if stunned, not so much by the blow as by sheer bewilderment caused by a happening which he believed impossible.

The roll lay at the tips of his forepaws, where it had dropped after hitting him between the eyes. He gave the roll a short glance, as though he thought it might be a live thing which might jump up and prove to him that he had been mistaken in believing his friend capable of doing such a thoughtless thing.

Now, raising his eyes from the roll, he let them wander across the floor and slowly up the counter until they ultimately came to rest on the Frenchman's frozen face. There they fixed themselves, as if held by magnetic power. No accusation was in the dog's eyes, simply a profound sadness, the sadness of one who infinitely trusts

[2]**sufficed** (sə·fīsd′): been enough.

somebody's honest friendship and then unexpectedly finds himself betrayed and cannot tell why it all happened.

The Frenchman, suddenly realizing what he had done, stood aghast as though he, by accident, had killed a human being. An involuntary shiver straightened his whole body, and he came to.

For a few seconds he stared at the dog with astonished eyes, as if seeing an apparition. Now the dog slowly rose, shook his head so that his ears flapped against his jaws, turned around, and went his way.

On seeing the dog disappear from the door, the Frenchman got irritated, looked around vaguely as if to find something very quick, without for a moment knowing exactly what it was; and looking down, his eyes came to rest upon a patron sitting at the counter who was just sticking a fork into a steak that had been placed before him by a waitress. Deftly, the Frenchman snatched that steak from the platter of the astonished diner who, with an earsplitting yell, jumped from his seat and loudly and energetically protested against the outrageous violation of a citizen's constitutional right to eat his food in peace.

Holding that steak between his fingers, the Frenchman darted out of the door and ran along the street. He saw the dog already trotting along in the next block, and he started running after him, whistling and calling, not in the least minding the people on the street who stopped to watch a lunatic with a steak dangling from his fingers whistling for a street dog to come and get it. When he reached the third block and could no longer see the dog or make out where he had gone, he let the steak drop and walked back to his restaurant.

"Perdone me, señor, siento mucho,"[3] he said to the patron, who in the meantime had returned to the counter where he had been given another steak in order to calm him. "Excuse me, friend, the steak wasn't good anyway, you see . . . besides, I just wanted to give it to somebody who might need it more than you! Please forget it, *amigo.* Order anything special, and whatever it may be, it'll be on the house."

"Caramba, such things don't happen in an ordinary day's work, so it's okay by me. Don't mention it. I got another steak, thanks to the elegant service you give in this here joint. And say, *señor,* as to that special order on the house, make mine a double-

[3]*Perdone . . . mucho:* Pardon me, sir; I'm very sorry.

sized pie à la mode, see. Okay?"

"Right, mister, and a double-sized à la mode it'll be, and you're welcome to it."

Restlessly walking about the place, moving in a daze, here pushing a chair, there a table, into another position, pulling a tablecloth straight, he presently reached that dark corner where the waitress was crouched, still crying.

"It's all right, Bertha; you can stay on. It wasn't all your fault. I'll murder that baker some day even if I get shot for it. Well, anyhow, I'll change that baker instead of you. That's all. I just got upset with that guy hollering about his cracked false tooth like a monkey gone mad."

"Thanks, boss," Bertha sniffled. "I'm really grateful, really I am, 'n sure thing I'll make it up to you by being a real help around here. See, boss, I got a mother and two brats to care for, and I tell you it ain't so easy to land another job as good as the one I got here, tips and all. . . ."

"Stop talking and get to work. I told you it's okay, so what are you kicking about?"

"I'm not kicking, not at all. I only meant to say thanks, boss." Turning to an impatient patron, she said, "Yeah, yeah, mister, I heard you the first time. Just keep your shirt on and wait a second, okay?"

The Frenchman consoled himself that the dog would come again next day. He wouldn't miss his lunch for a little misunderstanding like that. Why, things like that can happen every day in a dozen places, most anywhere. Every dog is whipped by his master now and then if he deserves it, and the dog stays on, forgetting all about it in an hour or so, perhaps. Dogs are faithful; they stick to the one who feeds them.

Yet despite reassuring himself, he couldn't feel at ease. For the rest of the day his conscience kept reminding him of the dog's habitual and quizzical grin which had amused him so much whenever he saw it, and he thought how it had died on the dog's face and changed into a desolate sadness, as though in that bad moment something had broken within the dog. The more he tried to forget the dog, the more frequently did he shoot a glance at the door, expecting to see him sitting there. Over and over he told himself that this dog was just a common mongrel, an ordinary street dog, living off the garbage cans, with no character or personality in particular. Yes, hand him a bone and you're his eternal friend. "Gosh," he finally muttered, "can I never concentrate on anything else but that stray dog which is not even mine? Why, I don't even know his name. Well, well. Let's forget it. He isn't worth the trouble, anyhow."

Next day, however, by three in the afternoon, the Frenchman had a good piece of steak ready, juicy and rare, with which to welcome the dog and to apologize at the same time for the insult he had given, and so renew the old friendship.

Half past three. And as if materialized by the striking of the nearest clock, there was the dog sitting at his usual place near the door.

The Frenchman's heart leaped with joy. "I knew it; I knew he'd be there," he said to himself with a satisfied smile.

Yet while saying so he felt slightly disappointed that this dog should prove to be exactly like any other dog. As he had come to like the dog, if not to love him, he had expected him to be different from other dogs, more proud or distinguished. Anyhow, he was pleased that the dog had come again and had given him a chance to make the animal feel that only by mistake had he been ill-treated, and that he, Monsieur Leblanc, had never meant to hurt him at all. Generously he forgave the dog his apparent lack of pride, telling himself that man must accept dogs as they are made, since man has no power whatever to change either their physical make-up or their canine nature.

There the dog sat, almost motionless, looking at the Frenchman with its velvety warm eyes as though searching that man's mind.

Greeting the animal with a wide-open smile of welcome, the Frenchman expected the dog to wag his tail and put on his clownish grin as a sign that no bad feelings existed between them. Yet the dog kept his mouth closed and made not the slightest move, either with his head or with his tail, even when he saw the Frenchman pick up the steak and wave it at him from behind the counter. The Frenchman jerked his head to indicate to the dog that he might come in and eat his lunch close to the counter, and thus feel more at home.

The dog, however, remained quietly sitting at his usual place outside the door, looking in at the Frenchman without a flicker of the eye, staring him straight in the face as if he meant to hypnotize the man.

Once more the Frenchman beckoned, with the steak swinging from between his fingers. He smacked his lips at the dog and clucked struuush-struuush to arouse the dog's appetite and to make his mouth water. This time the dog answered by slightly moving his tail, but stopped abruptly when, so it seemed, he realized what he was doing.

Presently the Frenchman, carrying the steak between his fingers, went out to the dog. Going close to him, he played the meat before the dog's nose, as he often used to do when he was in a mood to tease the dog a bit before letting him have the meal.

The dog, on seeing the Frenchman step up to him, raised his eyes, but otherwise did not move. When the dog refused the offered gift even as it dangled before his nose, the Frenchman, not losing his patience, laid the steak down on the pavement in front of the dog's forepaws. He stroked the animal, pulled one ear gently, and patted him on the back while the dog wagged his tail now — but so slightly that the move was barely noticeable. Yet no matter what the Frenchman did, the dog never let his eyes waver from the man's.

Nodding at the dog, with a big smile on his face, the Frenchman returned to his place behind the counter and watched the dog from there, expecting that he would now pick up the steak and eat it as usual out on the sidewalk.

However, the dog didn't. He lowered his head, sniffed without interest the meat before him on the ground, raised his eyes again to the man's face, stood up, turned around, and left.

The Frenchman rushed out onto the sidewalk and saw the dog ambling alongside the buildings, with not one look back at the man staring after him. Soon he disappeared amidst the people walking along the street.

Next day, punctual as ever, the dog again was sitting by the door, gazing at the face of his lost friend. And again, just as on the previous day, when the Frenchman approached, with a huge calf's bone this time, the dog only stared at him without taking the slightest interest in the man's gift laid before him on the sidewalk.

The dog, not for a moment ceasing to stare into the man's eyes, now wagged his tail when the Frenchman stroked and caressed the animal's head and fondled his ears.

Then the dog rose, pushed the man's caressing hand from his head with his nose, licked that hand over and over again for a full minute, once more looked up into the man's eyes — and without picking up the meaty bone or even sniffing at it, turned around, walked from the doorway, and left.

This was the last time the Frenchman ever saw that dog. He never came back to the café, and never again was he seen anywhere in the neighborhood.

My Elder Brother

LEO TOLSTOY

I was only a year and some months younger than Volodya; we grew up, studied, and played together. No distinction of elder and younger was made between us, but just about the time I am speaking of I began to realize that I was no companion for him, either in age, in interests, or in ability. It even seemed to me that Volodya himself was aware of his superiority and was proud of it. This idea (it may have been a wrong one) was inspired by my vanity — which suffered every time I came in contact with him. He was better than I in everything: at games, at lessons, in arguments, and in manners, and all this estranged me from him and occasioned me moral anguish which I could not understand. If I had said frankly when Volodya was given tucked linen shirts for the first time that I was vexed at not having shirts like that, I am sure I should have felt happier and not thought every time he arranged his collar that it was only done to annoy me.

What tormented me most was that it sometimes seemed to me Volodya understood what was going on inside me but tried to hide this.

Who has not noticed those mysterious unspoken relations which manifest themselves in a barely perceptible smile, a gesture, or a look between people who live together — brothers, friends, husband and wife, master and servant — especially when they do not cultivate mutual frankness in all things? How many unuttered desires, thoughts, and fears of being understood are expressed in one casual glance, when eyes meet shyly and hesitantly!

But perhaps my inordinate sensitiveness and tendency to analyze deceived me in this case. It may be Volodya did not feel at all as I did. He was impulsive, candid, and fickle in his enthusiasms. Carried away by the most diverse interests, he flung himself into them heart and soul.

He would suddenly conceive a passion for pictures, himself take up painting, spend all his money buying them, and beg them of his drawing master, of papa, and of grandmamma. Then it would be a rage for curios with which to adorn his table, collecting them from every room in the house; or a mania for novels, which he obtained on the sly and read all day and all night. . . . I could not help being enticed[1] by his hobbies, but I was too proud to imitate him and too young and not independent enough to choose a line for myself. But there was nothing I envied so much as Volodya's happy big-hearted disposition, which showed itself most strikingly when we quarreled. I always felt that he was behaving well, but I could not do likewise.

Once when his passion for ornaments was at its height, I went up to his table and there accidentally broke an empty brightly colored little scent bottle.

"Who asked you to touch my things?" demanded Volodya, coming into the room and seeing how I had upset the symmetry of the different treasures on his table. "And where is the scent bottle? You must have. . . . "

"I knocked it over by accident and it broke. What does it matter?"

"Do me the favor — never *dare* touch my things again," he said, putting the pieces of the broken flask together and looking at them sorrowfully.

"And you please don't issue orders," I retorted, "that's all. What's there to talk about in that?"

And I smiled, though I did not feel in the least like smiling.

"Yes, it's nothing to you, but it does matter to me," pursued Volodya, jerking his shoulder, a gesture he had inherited from papa. "He goes and breaks it, and then laughs, the nasty little *brat*!"

[1]**enticed** (en·tisd'): attracted; tempted.

"I'm a little brat; and you're big but you're stupid."

"I am not going to quarrel with you," said Volodya, giving me a slight push. "Go away."

"Don't push!"

"Get away!"

"Don't push, I tell you!"

Volodya took my arm and tried to drag me away from the table, but I was beside myself now; I got hold of the leg of the table and tipped it over. "There now!" And all his china and glass ornaments crashed to the floor.

"You disgusting little boy!" cried Volodya, trying to save some of his falling treasures.

"Well, now it is all over between us," I thought as I left the room. "We have quarreled for good."

We did not speak to each other till evening. I felt myself in the wrong and was afraid to look at him and could not settle to do anything all day. Volodya, on the contrary, did his lessons well, and after dinner talked and laughed with the girls as usual.

As soon as afternoon lessons were over, I left the room. I was too scared and uncomfortable and ashamed to be alone with my brother. After our history lesson in the evening I took my exercise books and started towards the door. As I passed Volodya, though I wanted to go up to him and make friends, I scowled and put on an angry expression. At that moment Volodya raised his head and with a faintly perceptible good-naturedly derisive[2] smile looked me full in the face. Our eyes met and I knew that he understood me, and knew that I knew that he understood me; but some irresistible feeling made me turn away.

"Nicky!" he said in a most natural voice without a scrap of pathos.[3] "Don't be cross any more. Forgive me if I offended you."

And he held out his hand.

Something that welled higher and higher seemed to be pressing my chest and hindering my breathing, but this only lasted a second; tears came to my eyes, and I felt better.

"Forgive . . . m-me, Vol-dya," I stammered, squeezing his hand. Volodya looked at me as if he could not make out at all why there should be tears in my eyes.

[2]**derisive** (di·rī'siv): mocking.
[3]**pathos** (pā'thos): pity; sympathy.

HOW THE SN

JULIUS LESTER

When the Lord made the Snake, he made him all the pretty colors he could think of — reds and browns and oranges — and the Lord put the Snake down here to decorate the ground and the bushes, to add a little color to everything.

Well, the Snake didn't mind ornamenting the earth, but he wondered if the Lord knew what a hard life he had to lead. And it was hard! Wasn't no doubt about that. He didn't have wings like the birds, so he couldn't fly. He didn't have fins, so he couldn't go for a swim in the river. He didn't have feet, so he couldn't run fast. All he could do was crawl in the dust. He didn't even mind that so much, but he couldn't figure out why the Lord had given him such poor eyesight. Man, the Snake was so blind he couldn't even see a flashlight if it was shining right in his eyes. He was so blind he couldn't see his hands in front of his face if he had had hands to put in front of his face. His eyesight was so bad he had to smell his way around. And to tell the truth, his nose was none too good, either.

Now you can imagine what it must've been like when the Snake had to go to the store or down to the laundromat. Crawling around in the dust like that and being half blind, he was always getting stepped on. He couldn't see anybody coming, and since he was down on the ground, nobody could see him. The other animals didn't mean to step on him, but you know how it is sometimes. You're in a hurry trying to get somewhere, and you just ain't got the time to see whether there's a blind snake with a no-smelling nose in your path. Some days, the mere thought of going somewhere tired the Snake out so much that he just didn't even bother getting up.

AKE GOT HIS
RATTLES

Mrs. Snake wasn't too happy about the situation, either. She couldn't even take the little snakes down to the playground anymore. The other children thought they were pretty jump ropes. So she was stuck in the house with the kids all day, and they were about to drive her out of her mind. "Why don't you do something?" she screamed at him one morning. "I went out to hang up the laundry this morning, and I almost got run down by two buffalo, three antelope, and a rabbit. I'm just sick and tired of this, and if you're half a snake at all, you'll do something about it."

Mr. Snake sighed. He'd heard it all so many times before. "Well, just what do you expect me to do? Grow arms? The Lord made us this way and ain't a thing we can do about it."

"Says who? You know as well as I do that God don't know what he's doing half the time. He just sits up there experimenting with this and that, and we were one of the experiments that didn't turn out well. You're going to do something, or I'm going to know the reason why. You're triflin', lazy, and no 'count. My mama told me not to marry you, and I should've listened to her!" And Mrs. Snake proceeded to call him a whole bunch of dirty names. She even jumped on back in his family and started playing the Dozens.[1] Said his mama was a lizard and his daddy was a fishing worm.

Well, Mr. Snake was trying to figure what number he should play that day, and he couldn't half concentrate with all the yelling his wife was doing, so he decided to crawl off to Heaven and talk it over with the Lord. He figured that the Lord wouldn't be able to do anything, but at least it would shut his wife up if she got the word directly from God.

[1]**playing the Dozens:** saying unfavorable, untrue things about his parents.

About two o'clock that afternoon, the Snake got up to Heaven. The Lord was sitting back in his great rocking chair reading *TV Guide.* "Well, what's going on, Mr. Snake?"

"Ain't nothing happening, Lord. Same ol' same ol'. You know."

"Yeah, know what you mean. Pull up a chair. I was just sitting here seeing what was gon' be on television in 1970. Course, that's a long way off, but I figured I'd check it out now, just in case I wanted to take a vacation about then."

The Snake curled up in a chair.

"Care for a cigar, Mr. Snake?"

"Thank you, Lord, but don't believe I do."

"I've been thinking about cutting down myself. Mrs. God tells me I smoke too much. She says it makes my breath smell bad."

"Is that so?"

"That's what she claims, Mr. Snake, and I just can't understand it. I brush after every meal. Well, that's neither here nor there. What's on your mind? You looking mighty well."

"Well, Lord, I ain't *doing* too well."

"You don't say."

"It's the truth. I hate to come up here bothering you and all, knowing how busy you are, but my wife just wouldn't let me rest until I did something."

"Well, why don't you tell me about it?"

"It's like this, Lord. You know my eyes are kinda bad. Seems to run in my family on both sides of the house. And me being low on the ground like I am, I can't tell when folks are coming down the road. As a result, me and my family are always getting stepped on. And Lord, I just ache all over, and you know I got a whole lot of muscles that can ache. The strain has made my nerves so bad that the rain beating on the leaves makes me jumpy."

"You in bad shape, Mr. Snake."

"I done tried Compoz and everything else you can think of for my nerves, Lord, but don't nothing help. Some days I think about going outside the house and getting stepped on, and I just don't get out of bed."

The Lord lit one of his big cigars and blew smoke rings for a few minutes. "Well, I didn't mean for nothing like that to happen to you." He reached in his pocket and pulled out a small bottle. "Here. This is poison. You put this in your mouth, and give the rest to all your kinfolks. This will be for you to use to protect yourself with. Anybody step on you, you can bite 'em and stick a little poison into 'em."

"Thank you, Lord. I sure thank you."

"Don't mention it, Mr. Snake. Glad to help out when I can. If you have any more problems, you just let me know."

The Snake didn't waste a minute getting down the ladder and going home. When he told his wife about the poison, she was so happy she kissed him. He told her that she best be cool and went to bed.

Well, a few days later, the Rabbit sent word through the forest that the animals had to send their union representative to a meeting. All the animals, that is, except the Snake.

"The meeting will now come to order!" the Rabbit announced. "Hey, Mr. Elephant! Can't you find a seat somewhere and stop blocking the sun? I don't know why the Lord made the dumbest thing in the forest the biggest."

The Elephant sat down.

"You still blocking the sunlight, Mr. Elephant."

"I'm sorry, Mr. Rabbit."

"Well, I understand that it ain't all your fault. You think you can hold your head to one side so a little light can come through? That's good! Now, everybody know why I called this meeting. It's that Snake! We got to do something, and do it quick, before he kills us all off. I've been to thirteen funerals this week, and today's only Tuesday."

"I know what you mean," the Frog put in. "That Snake killed three of my uncles, seven cousins, two aunts, and my brother-in-law. I have to admit, though, that getting rid of my brother-in-law was kind of a blessing."

"Well, blessing or no blessing, that Snake has been a terror ever since the Lord gave him that poison. He's been biting everything that shakes the bush. He can't see, but ain't a thing wrong with his hearing. Why, if he just hear any little thing, he's dead on it."

"And he's got the best aim I've ever seen," the Horse added.

"Who you telling?" exclaimed the Rabbit. "I saw a leaf shake on a bush, and I guess you know that's one dead leaf now. Snake put all sorts of poison in that leaf."

"Well, what we gon' do about it?" the Fox wanted to know.

"Tell him to move on over or we'll move on over him," said the Panther. "Tell him the next time he lays his forked tongue on anybody, we're going to off him."

"And suppose he lays his forked tongue on somebody and you *don't* off him," said the Rabbit.

"That's right," the Owl put in. "Hustlers don't call show-downs."

The Elephant said, "We could talk to him real nice-like and let him know that we don't mean him no harm, and all we want is our rights. He should be able to understand that."

"Like I said before, how come the Lord made you so big and gave you such a little brain? If your brain was magnified a hundred times, it would fit into the navel of a gnat and rattle around like a BB in a cornflakes box. Understand! He don't have to understand nothing he don't want to, as long as he's got so much power."

They talked and talked and talked and talked until the Rabbit got so disgusted he just took off for Heaven by himself. They were going to do all that talking and end up saying they should send somebody to talk to the Lord. The Rabbit wondered if he was the only one of the animals that had any sense.

The Lord was reading the newspaper when the Rabbit came up on the porch. "Yeah, Mr. Rabbit. Just between you and me, I ain't gon' be nowhere to be found between 1960 and 1970. And if you take my word, you'll go somewhere and hide yourself. That's going to be a rough ten years. Yes, yes. I ain't answering no prayers during that time. No, indeed!"

"Lord! What're you talking about?"

"You just remember what I said. O.K.?" He folded the newspaper and put it beside his chair. "Now, what's on your mind?"

"It's that Snake!"

"He was just up here a few days ago."

"Don't you think we know it. You gave him some poison, and he's bitten so many animals that a lot of them are packing up and going North. People are scared to let their wives and daughters go out alone at night. That's the truth! Lord, that Snake has bitten three hundred and thirty-seven animals, five oak trees, seventeen palm trees, and one stickerbush. Ha! Ha! Ha! You should've seen him bite into that stickerbush. Ha! Ha! Ha! His wife was pulling stickers off his lip for two hours. Served him right, the low-down—"

"I get the point. You go on back and tell Mr. Snake to get himself up here in a hurry."

"Well, it kind of dangerous to get too close to him, Lord."

"You tell the Elephant to yell the message in his general direction. Ain't a thing wrong with Mr. Snake's hearing."

A few hours later, the Snake was curled up in a chair next to the Lord.

"Lord, I sure want to thank you for that poison you gave me a few days ago. I haven't gotten stepped on since."

"That's just what I wanted to talk to you about. What's this I hear about you using that poison for insurance instead of protection?"

"Well, Lord, you know my eyes ain't too good, and I can't see who's my friend and who come to do me harm. So I just bite everybody that come along, and that way I'm always on the safe side. I've been stepped on so much that I just can't take any chances."

"I can understand that, Mr. Snake. But now you just about terrorized everybody."

"I didn't mean to, Lord. I sure didn't intend to do that."

The Lord reached in his pocket. "Here. You take these rattles and put 'em on your tail. When you hear something, shake your tail. That'll be a warning. If it's your friend, he'll stop and pass the time of day with you. And if it's your enemy, he'll just keep coming, and after that, it's you and him. You understand?"

"Yes, Lord. I do. And I thank you. It was getting mighty tiresome biting everything that came along. Mighty tiresome."

And that's how the Snake got his rattles. By that time, though, almost everybody was so afraid of him anyway that hardly anyone ever came around to see him. When they did, though, he shook his tail right hard, and it rattled through the forest, letting everybody know: this is Mr. Snake here. You can't step on me now.

Where?

KENNETH PATCHEN

There's a place the man always say
Come in here, child
No cause you should weep
Wolf never catch the rabbit
Golden hair never turn white with grief
Come in here, child
No cause you should moan
Brother never hurt his brother
Nobody here ever wander without a home
There must be some such place somewhere
But I never heard of it

Secrets are not meant to be learned.

The Wishing Well

by

PHILLIP BONOSKY

Beezie went home that night after listening to the caddy and lay awake thinking of the bottom of the well building higher and higher with dimes, nickels, quarters, and half dollars. That had been going on for years, the caddy had guessed; for years — the guys and girls would throw money into the water and make wishes.

"Will you come with me?" he said to Kozik. "Will you come? Just watch, I'll go down. But will you come?"

"If they catch us?"

"Kick us in the pants and let us go. We're just kids," he explained.

"How can we get in?"

"I'll find a way, but will you come?"

Kozik's yellow eyes considered. Over them his lids fell and opened like the lazy wings of a moth.

"How much will you give me?"

"A third," Beezie said.

"No, a half."

"You'll just watch. *I'll* go down the well!"

Kozik stared away.

"How much you think is down there?"

"A hundred dollars!"

"Gee!" His lids opened wide. "A hundred dollars!" he echoed. "You're crazy! Who'd let all that money lay down that old well? *Somebody* would take it! You're crazy!"

"You don't *know!*" Beezie replied. "You don't know *them!* They're rich!"

They pooled their money and took two streetcars — the first one carried them out of town; the second one, ten miles into the country, through the little hamlets, until up on a green hill, like a pearl, sat the big white house, partly screened by elm trees. It was late afternoon, and they were hungry but had no money to buy food. They ate sour-grass. They had no money either to ride back home again in case they failed. They hadn't thought about that.

"We'll wait until it gets dark," Beezie said.

"Then the fellas come down," Kozik protested.

"That's why you gotta watch. They'd see us in the daytime."

"How about dogs?"

"Dogs?"

"Yeah, how about dogs?"

But Beezie didn't answer.

They sat on a rock at the point where the driveway divided from the road. They stared solemnly as the bright sport cars loaded with strange women and stranger-looking men drove up. The cars were jagged with tennis rackets, golf clubs, polo sticks — the unrecognizable tools of their living. The women's hair didn't look like hair to the boys, and the hair on the men was odd too. They wore sunglasses — tortoise shelled or pink coral, in every shape. They also spoke differently, and though the boys heard whole phrases, they understood nothing that was said.

They watched them like exotic birds from a green jungle.

"Besides," Beezie said, remembering the final proof, "it was in the papers — about their doing it; the wishing well, throwing money in it."

"How you going to find where it is?"

"Down there," Beezie said mysteriously, pointing to a clump of trees down the hill from the house. "Down there. That's where."

They were both hungry but didn't realize it. Their hunger was trapped inside their excitement. They lay beside the road, munching sour-grass or chewing orchard grass; they picked off the green haws[1] from the bushes and threw them at ants and at birds. Cars came and went, and each time one arrived, both fell silent and stared at its strange cargo.

"Suppose they catch us?" Kozik echoed.

It was still less than dark when, worn out with waiting, Beezie cried: "Let's start!"

"They'll see us!"

"No," he fired back. "Hear the music!"

They stood to listen. From the house on the hill came music. Lights had begun to shine, too. But the music, most of all, came; it made the house seem like a handful of another world set down on top of the hill in the middle of the woods and fields, being by itself.

"They're listening to that. They won't hear us."

There was a wire fence running along the road. It was tipped with sharpened, turned-over naked ends. When they touched it with their hands, they jumped back.

"Oh!" Kozik cried, backing away from it and staring at it with horror. His lids fell heavily and then opened.

From behind the charged fence the music came. Beezie's hands tingled and his scalp had grown cold. It seemed to him for a moment that they had taken the wrong streetcar and this was the wrong place, it wasn't even in the same country.

Away from them the fence crossed a creek, and between the creek and the bottom of the fence was a foot and a half of space. They stood above it debating silently for a long time before taking the risk of climbing under the fence. But they only wet their shoes.

Inside the fence it felt different. Now even the music seemed nearer and clearer. Beezie felt his heart pounding right under his shirt; he wanted to still it with his hand.

The music charged on them. They could distinguish a cymbal clash; even, for one brief moment that made them stand still with cold, the crystal laugh of a girl; then it was shut off.

The creek led through thicker and thicker bushes. Beezie could smell crushed blackberries. A bridle path[2] followed along, but they did not dare to walk the path; instead, they threaded themselves

[1]**haws:** hawthorn berries.
[2]**bridle path:** trail for horseback riding.

through the bushes. When they arrived, at first they didn't recognize the spot. The well was under a pergola,[3] over which grapes grew. There were stone seats around, off a little; inside the pergola was the stone well. It had an old wooden handle, an old green-mossy axle around which the rope wound which let down a wooden bucket into the well. A wooden lid with a hand-carved handle stood beside it, green with faint moss.

Kozik's moth-lids blinked. "How deep is it?" he asked.

Beezie looked at the well, then leaned over into it. He threw a pebble down and listened. A *plunk* followed, and he said, thoughtfully, "It's pretty far down, but not *deep,* I think."

He stood silent. Neither did Kozik speak. Fireflies suddenly swarmed out of the bushes and flew in and out. The music was faint again, but through the trees they could see the lighted windows; then the music, flung at them suddenly, crashed through and jarred him.

"Well," Beezie said, remotely, "I'll go down. You unwind it, slow."

"All right."

"And watch."

"All right."

"When I'm down, you put in the peg in there." He showed how to peg the winder so that it would no longer turn.

He took off his shoes and stockings and rolled up his trouser legs above his knees. The moonlight made his legs look pale. He climbed into the bucket. His white face hovered over the rim of the well for a moment and, hushed, he said: "We'll be rich, Kozik!"

"We'll count it when we get home," Kozik said.

"Yes," Beezie answered. Kozik began to unwind. His white face staring at him, Beezie disappeared.

The wood creaked loud in the stillness. It unwound slowly, and the bucket bumped solemnly from side to side of the well. Looking up, he could see only gray, and with an upward leap of his heart, a single shining star. Kozik he could not see.

If the rope broke, he thought; but he gripped it and felt its strength. Suddenly the unwinding stopped and he dangled in the middle of the well. Looking up, he saw silhouettes. Then heard

[3]**pergola** (pėr′gə·lə): roof supported by posts, over which plants grow.

voices. Suddenly money began to fall past him, plunking into the water; there was laughter upstairs. "I was right," he whispered to himself.

He sat twirling slowly, lazily, round and round. As he came near the wall of the well, he gently kicked it and swung away. The wall was slimy and cold to the touch of his bare toes; the well smelled of moss and frogs and salamanders.

Then the rope began to unwind again, its creaking sounding wooden and old, like an ungreased wheel. Abruptly he landed on the water and cried out: "Down!" and the creaking stopped. Kozik leaned his head over and cried: "Down?"

"Down," he answered.

He dangled his feet over into the water and began to reach through it for footing. The water was so cold he gasped. He hunted with his foot but found nothing to stand on. "Hell," he said to himself.

Now he took his trousers off and let himself over the rim of the bucket into the water, shuddering as he sank, and hanging onto the bucket stretched his feet as far as they could go. Still there was no footing. Now, closing his eyes, he let go of the bucket and sank down over his head. Touching bottom, he fell to his knees and began to grope along the pebbled bottom picking up coins. Then he shot to the water top, and his lungs beating, and his teeth beginning to chatter slightly, he dumped the coins into the bucket. Again he let go of it and sank to the bottom; underneath he worked furiously, with slow motions, as in a dream, holding his breath until his head blew up with blood; then he burst to the surface again, and again dumped the coins into the wooden bucket. And again he sank down.

He seemed to rise out of the water like an icicle. His teeth were clenched in his frozen mouth. It was too dark to see what coins he had brought up; his head was so filled with cold he could not hear what Kozik shouted down to him. He heard a cry, but dived down again to the bottom of the well, scratched desperately, then, both fists clutched with coins, rose again to the surface.

He reached for the bucket — treaded water reaching for it, his eyes still closed. Then, looking up suddenly, he saw it whirling slowly upward, and beyond it, over the rim of the well, the white face of a girl looking down.

Treading water, he only watched it, unable to think. His head was growing cold from the inside. His feet seemed to tread as if forgotten by him. Then he heard them.

"Look at that!" the man cried. "Money!"

"Oh," the girl said.

"It must have fallen into the bucket instead," the man said.

"This is our lucky night!" the girl cried, laughing.

"Throw it back in," the man ordered.

"No," she said. "If we were smart enough to bring the bucket up — we *win!*" she cried.

"Some of it, anyhow," the man said.

Money spattered down beside him, hitting him, falling around him.

He suddenly began to sink. Loosening one of his fists spasmodically[4] and letting the coins sink, he flung a hand toward the wall and there caught onto a crack filled with moss. Holding onto it, he rested. He smelled worms in the well.

Now the cold grew in his body; he felt it touch his bones, get into his stomach, encase his lungs with a film of ice. Like a sound over mountains, he felt the first pangs of cramps gathering in his loins, begin slightly to clutch at his stomach. "Let them go away," he said.

In his other fist the coins felt jagged. His hand was shut so tightly over them it seemed to be locked. Trying to open it slightly, he discovered that he could not. He laughed at it, surprised, at his fist.

Above him they were still talking; no longer arguing, they were talking to each other. Where was Kozik? *Where are you, Kozik?* he cried. Did he run home, did he get scared and run home? *No money for you, Kozik,* he cried; *no money for you, you lost your share for running away! You're yellow; you won't get any at all. I'll keep it; I'll learn you to be yellow, Kozik,* he cried. *You'd better come back!*

He felt that his fingers holding the crack were losing sensation. Now, remotely, he felt his legs again. They were still treading, had been treading all the time. He felt grateful to them that they were still treading, because at this point he felt that they didn't need to if they didn't want to. Parts of his body became objectified: his fist

[4]**spasmodically** (spaz·mod′ik·lē): suddenly; uncontrollably.

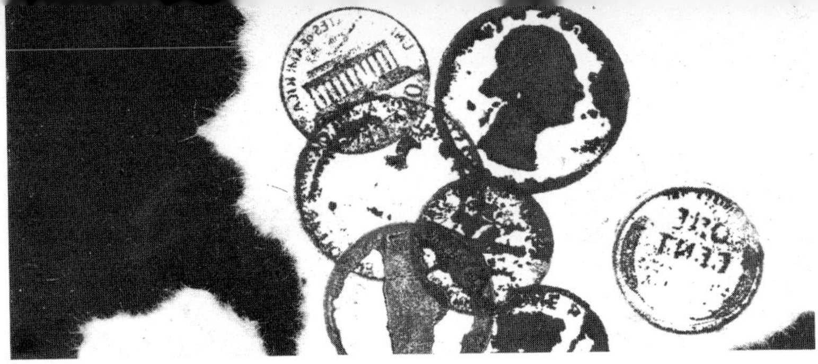

was one, his legs were another, his free hand holding to the crack was another, his stomach was another; *himself* was in his head. He promised the parts things and chided[5] them too; he also reasoned with them. "Hold on tighter," he admonished his hand. To his fist he said: "You hurt."

The water was up to his neck. He had left his shirt on and it was soaked. Suddenly he thought of his pants. Did they throw them away? He cried, "My pants!" He was cold now; it would be cold; how could he go anywhere without pants? He wanted to laugh, thinking of himself going home without pants; then his father whipping him — he wanted to cry now; then suddenly his body began to ache from head to foot. His head began to give in; it began to grow colder and colder and throb, it began to collect cold behind his eyes, at the back of his nose. All over it began to attack at once, and he saw himself shutting his eyes and slipping swiftly to the bottom of the well.

"Oh, Kozik!" he cried. "Kozik!"

He lifted his head to stare at the top of the well. The open circle of the well was gone. He couldn't see the star! They had put the lid over him! He was dead. *I'm in my grave without a coffin,* he said.

Suddenly he began to scream, pulling himself out of the water and falling back again. His feet thrashed beneath him as he screamed. His free hand searched the entire slimy wall for a place to put his feet.

Then he heard the clunk of the bucket against the side of the well. He saw a rim of light, a nimbus, around the bucket; then he caught a glimpse of the star again. It was the bucket coming down. Suddenly he began to sob for it. *Come down, come down,* he cried. *Oh, come down, come down!*

It hit beside him, tilting in the water. It was too far away to reach with his fist; the fist was clenched shut, it could not open. The coins were jagged inside. He let go of the crack with his other hand

[5]**chided** (chid'id): scolded.

and fell forward into the water, catching the bucket. It swung toward him and hit his nose, bringing cold tears. He pulled his arm over it, pulled himself half over on it, and lay still. In his mouth he felt the loud chatter of his teeth; they went so fast they began to hurt.

Now, slowly, the bucket began to rise. He lay half across it, his legs dangling, his head pointed downward, his eyes closed. He felt the thick rope throb against his neck. Then he was there; he felt hands, heard a voice.

Kozik struggled with him, finally pulled him off the bucket and fell to the ground with him. He was crying, "Beezie, what's the matter, Beezie? Can't you talk, Beezie? They didn't go away! They kept talking and talking, Beezie, and kissing — they wouldn't go away! They kept kissing, Beezie!"

He looked down; he stared into the blue face of his friend and cried: "Beezie, God, Beezie! You look sick! What's the matter, Beezie; can't you talk?"

Beezie rolled over face down onto the grass. He gripped his teeth into the grass and bit. His shoulders began to shake. The other started at him with horror and fear. "Beezie!" he kept crying. "It's not my fault, Beezie! It's not my fault! I didn't say to come out here!" Then, suddenly, freshly: "They threw your pants away and I saved them!"

Bit by bit the quaking died down; he felt his separate feet and hands and stomach come together again. The ache subsided behind his eyes; the pain slowly dwindled.

The first thing he said: "Did you get the pants, you said?"

"Yes," the other replied, holding them up. "They're not even wet!"

Beezie stood up and put his trousers on. His fist was still closed. "Let's go home," he said.

They began to walk through the woods; suddenly they began to run. When they came to the fence, they crawled under it, drenching themselves; and on the other side they kept running until they reached the car stop.

Beezie stretched his fist over to Kozik.

"Open it," he ordered.

Finger by finger Kozik pulled the paralyzed fingers apart and unfolded the fist. In it lay four dimes and two quarters. They both stared at it for a long time.

Slowly Beezie took one of the quarters and gave it to Kozik.

"This is your share," he said.

The other looked at the quarter. Then, taking a dime, Beezie said, "This is because you didn't run away." He gave it to him.

As they waited for the car to take them home, Kozik asked: "Shall we come back again?"

Beezie thought a long time before he finally answered.

"No," he said.

JUG OF SILVER

TRUMAN CAPOTE

After school I used to work in the Valhalla drugstore. It was owned by my uncle, Mr. Ed Marshall. I call him Mr. Marshall because everybody, including his wife, called him Mr. Marshall. Nevertheless he was a nice man.

This drugstore was maybe old-fashioned, but it was large and dark and cool: during summer months there was no pleasanter place in town. At the left, as you entered, was a tobacco-magazine counter behind which, as a rule, sat Mr. Marshall: a squat, square-faced, pinkfleshed man with looping, manly, white mustaches. Beyond this counter stood the beautiful soda fountain. It was very antique and made of fine, yellowed marble, smooth to the touch but without a trace of cheap glaze. Mr. Marshall bought it at an auction in New Orleans in 1910 and was plainly proud of it. When you sat on the high, delicate stools and looked across the fountain, you could see yourself reflected softly, as though by candlelight, in a row of ancient, mahogany-framed mirrors. All general merchandise was displayed in glass-doored, curio-like cabinets that were locked with brass keys. There was always in the air the smell of syrup and nutmeg and other delicacies.

The Valhalla was the gathering place of Wachata County till a certain Rufus McPherson came to town and opened a second drugstore directly across the courthouse square. This old Rufus McPherson was a villain; that is, he took away my uncle's trade. He

Capote: pronounced kə·pō′tē.

installed fancy equipment such as electric fans and colored lights; he provided curb service and made grilled-cheese sandwiches to order. Naturally, though some remained devoted to Mr. Marshall, most folks couldn't resist Rufus McPherson.

For a while, Mr. Marshall chose to ignore him: if you were to mention McPherson's name, he would sort of snort, finger his mustaches, and look the other way. But you could tell he was mad. And getting madder. Then one day toward the middle of October I strolled into the Valhalla to find him sitting at the fountain playing dominoes and drinking wine with Hamurabi.

Hamurabi was an Egyptian and some kind of dentist, though he didn't do much business as the people hereabouts have unusually strong teeth, due to an element in the water. He spent a great deal of his time loafing around the Valhalla and was my uncle's chief buddy. He was a handsome figure of a man, this Hamurabi, being dark-skinned and nearly seven feet tall; the matrons of the town kept their daughters under lock and key and gave him the eye themselves. He had no foreign accent whatsoever, and it was always my opinion that he wasn't any more Egyptian than the man in the moon.

Anyway, there they were swigging red Italian wine from a gallon jug. It was a troubling sight, for Mr. Marshall was a renowned teetotaler.[1] So naturally, I thought: Oh, golly, Rufus

[1]**renowned teetotaler:** person famous for avoiding alcoholic drinks.

McPherson has finally got his goat. That was not the case, however.

"Here, son," said Mr. Marshall, "come have a glass of wine."

"Sure," said Hamurabi, "help us finish it up. It's storebought, so we can't waste it."

Much later, when the jug was dry, Mr. Marshall picked it up and said, "Now we shall see!" And with that disappeared out into the afternoon.

"Where's he off to?" I asked.

"Ah," was all Hamurabi would say. He liked to devil me.

A half hour passed before my uncle returned. He was stooped and grunting under the load he carried. He set the jug atop the fountain and stepped back, smiling and rubbing his hands together. "Well, what do you think?"

"Ah," purred Hamurabi.

"Gee . . . ," I said.

It was the same wine jug, God knows, but there was a wonderful difference; for now it was crammed to the brim with nickels and dimes that shone dully through the thick glass.

"Pretty, eh?" said my uncle. "Had it done over at the First National. Couldn't get in anything bigger-sized than a nickel. Still, there's lotsa money in there, let me tell you."

"But what's the point, Mr. Marshall?" I said. "I mean, what's the idea?"

Mr. Marshall's smile deepened to a grin. "This here's a jug of silver, you might say. . . . "

"The pot at the end of the rainbow," interrupted Hamurabi.

" . . . and the idea, as you call it, is for folks to guess how much money is in there. For instance, say you buy a quarter's worth of stuff — well, then you get to take a chance. The more you buy, the more chances you get. And I'll keep all guesses in a ledger till Christmas Eve, at which time whoever comes closest to the right amount will get the whole shebang."

Hamurabi nodded solemnly. "He's playing Santa Claus — a mighty crafty Santa Claus," he said. "I'm going home and write a book: *The Skillful Murder of Rufus McPherson.*" To tell the truth, he sometimes did write stories and send them out to the magazines. They always came back.

It was surprising, really like a miracle, how Wachata County took to the jug. Why, the Valhalla hadn't done so much business since Station Master Tully, poor soul, went stark raving mad and claimed to have discovered oil back of the depot, causing the town to be overrun with wildcat prospectors. Even the poolhall bums who never spent a cent on anything not connected with whiskey or women took to investing their spare cash in milk shakes. A few elderly ladies publicly disapproved of Mr. Marshall's enterprise as a kind of gambling, but they didn't start any trouble, and some even found occasion to visit us and hazard a guess. The school kids were crazy about the whole thing, and I was very popular because they figured I knew the answer.

"I'll tell you why all this is," said Hamurabi, lighting one of the Egyptian cigarettes he bought by mail from a concern in New York City. "It's not for the reason you may imagine; not, in other words, avidity.[2] No. It's the mystery that's enchanting. Now you look at those nickels and dimes and what do you think: ah, so much! No, no. You think: ah, *how* much? And that's a profound question, indeed. It can mean different things to different people. Understand?"

And oh, was Rufus McPherson wild! When you're in trade, you count on Christmas to make up a large share of your yearly profit, and he was hard pressed to find a customer. So he tried to imitate the jug; but being such a stingy man he filled his with pennies. He also wrote a letter to the editor of *The Banner,* our weekly paper, in which he said that Mr. Marshall ought to be "tarred and feathered and strung up for turning innocent little children into confirmed gamblers and sending them down the path to Hell!" You can imagine what kind of laughingstock he was. Nobody had anything for McPherson but scorn. And so by the middle of November he just stood on the sidewalk outside his store and gazed bitterly at the festivities across the square.

At about this time Appleseed and sister made their first appearance.

He was a stranger in town. At least no one could recall ever having seen him before. He said he lived on a farm a mile past Indian Branches; told us his mother weighed only seventy-four pounds and that he had an older brother who would play the fiddle at anybody's wedding for fifty cents. He claimed that Appleseed was the only name he had and that he was twelve years old. But his sister, Middy, said he was eight. His hair was straight and dark

[2]**avidity** (ə·vid'ə·tē): *here,* greed.

yellow. He had a tight, weathertanned little face with anxious green eyes that had a very wise and knowing look. He was small and puny and high-strung; and he wore always the same outfit: a red sweater, blue denim britches, and a pair of man-sized boots that went clop-clop with every step.

It was raining that first time he came into the Valhalla; his hair was plastered round his head like a cap, and his boots were caked with red mud from the country roads. Middy trailed behind as he swaggered like a cowboy up to the fountain where I was wiping some glasses.

"I hear you folks got a bottle fulla money you fixin' to give 'way," he said, looking me square in the eye. "Seein' as you-all are givin' it away, we'd be obliged iffen you'd give it to us. Name's Appleseed, and this here's my sister, Middy."

Middy was a sad, sad-looking kid. She was a good bit taller and older-looking than her brother: a regular beanpole. She had tow-colored[3] hair that was chopped short, and a pale, pitiful little face. She wore a faded cotton dress that came way up above her bony knees. There was something wrong with her teeth, and she tried to conceal this by keeping her lips primly pursed like an old lady.

"Sorry," I said, "but you'll have to talk with Mr. Marshall."

So sure enough he did. I could hear my uncle explaining what he would have to do to win the jug. Appleseed listened attentively, nodding now and then. Presently he came back and stood in front of the jug, and touching it lightly with his hand, said, "Ain't it a pretty thing, Middy?"

Middy said, "Is they gonna give it to us?"

"Naw. What you gotta do, you gotta guess how much money's inside there. And you gotta buy two bits' worth so's even to get a chance."

[3]**tow-colored:** whitish-blonde, like flax.

"Huh, we ain't got no two bits. Where you 'spec we gonna get us two bits?"

Appleseed frowned and rubbed his chin. "That'll be the easy part; just leave it to me. The only worrisome thing is: I can't just take a chance and guess. . . . I gotta *know.*"

Well, a few days later they showed up again. Appleseed perched on a stool at the fountain and boldly asked for two glasses of water, one for him and one for Middy. It was on this occasion that he gave out the information about his family: ". . . then there's Papa Daddy, that's my mama's papa, who's a Cajun,[4] an' on accounta that he don't speak English good. My brother, the one what plays the fiddle, he's been in jail three times. . . . It's on accounta him we had to pick up and leave Louisiana. He cut a fella bad in a razor fight over a woman ten years older'n him. She had yellow hair."

Middy, lingering in the background, said nervously, "You oughtn't to be tellin' our personal private fam'ly business thataway, Appleseed."

"Hush now, Middy," he said, and she hushed. "She's a good little gal," he added, turning to pat her head, "but you can't let her get away with much. You go look at the picture books, honey, and stop frettin' with your teeth. Appleseed here's got some figurin' to do."

This figuring meant staring hard at the jug, as if his eyes were trying to eat it up. With his chin cupped in his hand, he studied it for a long period, not batting his eyelids once. "A lady in Louisiana told me I could see things other folks couldn't see 'cause I was born with a caul[5] on my head."

"It's a cinch you aren't going to see how much there is," I told him. "Why don't you just let a number pop into your head, and maybe that'll be the right one."

"Uh, uh," he said, "too darn risky. Me, I can't take no sucha chance. Now, the way I got it figured, there ain't but one sure-fire thing, and that's to count every nickel and dime."

"Count!"

"Count what?" asked Hamurabi, who had just moseyed inside and was settling himself at the fountain.

[4]**Cajun** (kā'jən): descendant of French people forced to leave the Nova Scotia area in the eighteenth century.

[5]**caul:** membrane (thin sheet of tissue).

"This kid says he's going to count how much is in the jug," I explained.

Hamurabi looked at Appleseed with interest. "How do you plan to do that, son?"

"Oh, by countin'," said Appleseed matter-of-factly.

Hamurabi laughed. "You better have X-ray eyes, son, that's all I can say."

"Oh, no. All you gotta do is be born with a caul on your head. A lady in Louisiana told me so. She was a witch; she loved me, and when my ma wouldn't give me to her, she put a hex[6] on her, and now my ma don't weigh but seventy-four pounds."

"Ve-ry in-ter-esting," was Hamurabi's comment as he gave Appleseed a queer glance.

Middy sauntered up, clutching a copy of *Screen Secrets*. She pointed out a certain photo to Appleseed and said: "Ain't she the

[6]**hex:** evil spell.

nicest-lookin' lady? Now you see, Appleseed, you see how pretty her teeth are? Not a one outa joint."

"Well, don't you fret none," he said.

After they left, Hamurabi ordered a bottle of orange Nehi and drank it slowly, while smoking a cigarette. "Do you think maybe that kid's okay upstairs?" he asked presently in a puzzled voice.

Small towns are best for spending Christmas, I think. They catch the mood quicker and change and come alive under its spell. By the first week in December house doors were decorated with wreaths, and store windows were flashy with red paper bells and snowflakes of glittering isinglass.[7] The kids hiked out into the woods and came back dragging spicy evergreen trees. Already the women were busy baking fruitcakes, unsealing jars of mincemeat, and opening bottles of blackberry and scuppernong wine. In the courthouse square a huge tree was trimmed with silver tinsel and colored electric bulbs that were lighted up at sunset. Late of an afternoon you could hear the choir in the Presbyterian church practicing carols for their annual pageant. All over town the japonicas[8] were in full bloom.

The only person who appeared not the least touched by this heartwarming atmosphere was Appleseed. He went about his declared business of counting the jug-money with great, persistent care. Every day now he came to the Valhalla and concentrated on the jug, scowling and mumbling to himself. At first we were all fascinated, but after a while, it got tiresome and nobody paid him any mind whatsoever. He never bought anything, apparently having never been able to raise the two bits. Sometimes he'd talk to Hamurabi, who had taken a tender interest in him and occasionally stood treat to a jawbreaker or a penny's worth of licorice.

"Do you still think he's nuts?" I asked.

"I'm not so sure," said Hamurabi. "But I'll let you know. He doesn't eat enough. I'm going to take him over to the Rainbow Café and buy him a plate of barbecue."

"He'd appreciate it more if you'd give him a quarter."

"No. A dish of barbecue is what he needs. Besides, it would be better if he never was to make a guess. A high-strung kid like that,

[7]**isinglass** (ī'zing·glas): clear, sheetlike mineral.
[8]**japonicas** (jə·pon'ə·kəz): red-flowering shrubs.

so unusual, I wouldn't want to be the one responsible if he lost. Say, it would be pitiful."

I'll admit that at the time Appleseed struck me as being just funny. Mr. Marshall felt sorry for him, and the kids tried to tease him but had to give it up when he refused to respond. There you could see him plain as day sitting at the fountain with his forehead puckered and his eyes fixed forever on that jug. Yet he was so withdrawn you sometimes had this awful creepy feeling that, well, maybe he didn't exist. And when you were pretty much convinced of this, he'd wake up and say something like, "You know, I hope a 1913 buffalo nickel's in there. A fella was tellin' me he saw where a 1913 buffalo nickel's worth fifty dollars." Or, "Middy's gonna be a big lady in the picture shows. They make lotsa money, the ladies in the picture shows do, and then we ain't gonna never eat another collard green as long as we live. Only Middy says she can't be in the picture shows 'less her teeth look good."

Middy didn't always tag along with her brother. On those occasions when she didn't come, Appleseed wasn't himself; he acted shy and left soon.

Hamurabi kept his promise and stood treat to a dish of barbecue at the café. "Mr. Hamurabi's nice, all right," said Appleseed afterward, "but he's got peculiar notions: has a notion that if he lived in this place named Egypt, he'd be a king or somethin'."

And Hamurabi said, "That kid has the most touching faith. It's a beautiful thing to see. But I'm beginning to despise the whole business." He gestured toward the jug. "Hope of this kind is a cruel thing to give anybody, and I'm damned sorry I was ever a party to it."

Around the Valhalla the most popular pastime was deciding what you would buy if you won the jug. Among those who participated were: Solomon Katz, Phoebe Jones, Carl Kuhnhardt, Puly Simmons, Addie Foxcroft, Marvin Finkle, Trudy Edwards, and Erskine Washington. And these were some of their answers: a trip to and a permanent wave in Birmingham, a second-hand piano, a Shetland pony, a gold bracelet, a set of *Rover Boys* books, and a life insurance policy.

Once Mr. Marshall asked Appleseed what he would get. "It's a secret," was the reply, and no amount of prying could make him tell. We took it for granted that whatever it was, he wanted it real bad.

Honest winter, as a rule, doesn't settle on our part of the country till late January, and then is mild, lasting only a short time. But in the year of which I write we were blessed with a singular cold spell the week before Christmas. Some still talk of it, for it was so terrible: water pipes froze solid; many folks had to spend the days in bed snuggled under their quilts, having neglected to lay in enough kindling for the fireplace; the sky turned that strange dull gray as it does just before a storm, and the sun was pale as a waning moon. There was a sharp wind: the old dried-up leaves of last fall fell on the icy ground, and the evergreen tree in the courthouse square was twice stripped of its Christmas finery. When you breathed, your breath made smoky clouds. Down by the silk mill where the very poor people lived, the families huddled together in the dark at night and told tales to keep their minds off the cold. Out in the country the farmers covered their delicate plants with gunny sacks and prayed; some took advantage of the weather to slaughter their hogs and bring the fresh sausage to town. Mr. R. C. Judkins, our town drunk, outfitted himself in a red cheesecloth suit and played Santa Claus at the five 'n' dime. Mr. R. C. Judkins was the father of a big family, so everybody was happy to see him sober enough to earn a dollar. There were several church socials, at one of which Mr. Marshall came face to face with Rufus McPherson: bitter words were passed but not a blow was struck.

Now, as has been mentioned, Appleseed lived on a farm a mile below Indian Branches; this would be approximately three miles from town; a mighty long and lonesome walk. Still, despite the cold, he came every day to the Valhalla and stayed till closing time which, as the days had grown short, was after nightfall. Once in a while he'd catch a ride partway home with the foreman from the silk mill, but not often. He looked tired, and there were worry lines about his mouth. He was always cold and shivered a lot. I don't think he wore any warm drawers underneath his red sweater and blue britches.

It was three days before Christmas when out of the clear sky, he announced: "Well, I'm finished. I mean I know how much is in the bottle." He claimed this with such grave, solemn sureness it was hard to doubt him.

"Why, say now, son, hold on," said Hamurabi, who was present. "You can't know anything of the sort. It's wrong to think so: you're just heading to get yourself hurt."

"You don't need to preach to me, Mr. Hamurabi. I know what I'm up to. A lady in Louisiana, she told me. . . . "

"Yes yes yes — but you got to forget that. If it were me, I'd go home and stay put and forget about this damned jug."

"My brother's gonna play the fiddle at a wedding over in Cherokee City tonight, and he's gonna give me the two bits," said Appleseed stubbornly. "Tomorrow I'll take my chance."

So the next day I felt kind of excited when Appleseed and Middy arrived. Sure enough, he had his quarter: it was tied for safekeeping in the corner of a red bandanna.

The two of them wandered hand in hand among the showcases, holding a whispery consultation as to what to purchase. They decided finally on a thimble-sized bottle of gardenia cologne which Middy promptly opened and partly emptied on her hair. "I smells like. . . . Oh, darlin', I ain't never smelled nothin' as sweet. Here, Appleseed, honey, let me douse some on your hair." But he wouldn't let her.

Mr. Marshall got out the ledger in which he kept his records, while Appleseed strolled over to the fountain and cupped the jug between his hands, stroking it gently. His eyes were bright and his cheeks flushed from excitement. Several persons who were in the drugstore at the moment crowded close. Middy stood in the background quietly scratching her leg and smelling the cologne. Hamurabi wasn't there.

Mr. Marshall licked the point of his pencil and smiled. "Okay, son, what do you say?"

Appleseed took a deep breath. "Seventy-seven dollars and thirty-five cents," he blurted.

In picking such an uneven sum he showed originality, for the run-of-the-mill guess was a plain round figure. Mr. Marshall repeated the amount solemnly as he copied it down.

"When'll I know if I won?"

"Christmas Eve," someone said.

"That's tomorrow, huh?"

"Why, so it is," said Mr. Marshall, not surprised. "Come at four o'clock."

During the night the thermometer dropped even lower, and toward dawn there was one of those swift, summerlike rainstorms, so that the following day was bright and frozen. The town was like a picture postcard of a Northern scene, what with icicles sparkling whitely on the trees and frost flowers coating all windowpanes. Mr. R. C. Judkins rose early and, for no clear reason, tramped the streets ringing a supper bell, stopping now and then to take a swig of whiskey from a pint which he kept in his hip pocket. As the day was windless, smoke climbed lazily from various chimneys straightway to the still, frozen sky. By mid-morning the Presbyterian choir was in full swing; and the town kids (wearing horror masks, as at Halloween) were chasing one another round and round the square, kicking up an awful fuss.

Hamurabi dropped by at noon to help us fix up the Valhalla. He brought along a fat sack of satsumas,[9] and together we ate every last one, tossing the hulls into a newly installed potbellied stove (a present from Mr. Marshall to himself) which stood in the middle of the room. Then my uncle took the jug off the fountain, polished and placed it on a prominently situated table. He was no help after that whatsoever, for he squatted in a chair and spent his time tying and retying a tacky green ribbon around the jug. So Hamurabi and I had the rest to do alone; we swept the floor and washed the mirrors and dusted the cabinets and strung streamers of red and green crepe paper from wall to wall. When we were finished, it looked very fine and elegant.

But Hamurabi gazed sadly at our work and said: "Well, I think I better be getting along now."

"Aren't you going to stay?" asked Mr. Marshall, shocked.

"No, oh, no," said Hamurabi, shaking his head slowly. "I don't want to see that kid's face. This is Christmas and I mean to have a rip-roaring time. And I couldn't, not with something like that on my conscience. Hell, I wouldn't sleep."

"Suit yourself," said Mr. Marshall. And he shrugged, but you could see he was really hurt. "Life's like that — and besides, who knows, he might win."

Hamurabi sighed gloomily. "What's his guess?"

"Seventy-seven dollars and thirty-five cents," I said.

"Now I ask you, isn't that fantastic?" said Hamurabi. He slumped in a chair next to Mr. Marshall and crossed his legs and lit a cigarette. "If you got any Baby Ruths, I think I'd like one; my mouth tastes sour."

[9]**satsumas** (sat'sü·məz): type of tangerine, first imported from China and now grown commercially in Florida and Alabama.

As the afternoon wore on, the three of us sat around the table feeling terribly blue. No one said hardly a word, and as the kids had deserted the square, the only sound was the clock tolling the hour in the courthouse steeple. The Valhalla was closed to business, but people kept passing by and peeking in the window. At three o'clock Mr. Marshall told me to unlock the door.

Within twenty minutes the place was jam full; everyone was wearing his Sunday best, and the air smelled sweet, for most of the little silk-mill girls had scented themselves with vanilla flavoring. They scrunched up against the walls, perched on the fountain, squeezed in wherever they could; soon the crowd had spread to the sidewalk and stretched into the road. The square was lined with team-drawn wagons and Model T Fords that had carted farmers and their families into town. There was much laughter and shouting and joking — several outraged ladies complained of the cursing and the rough, shoving ways of the younger men, but nobody left. Everybody was making the best of a good thing. It's usually so quiet around here: nothing much ever happens. It's safe to say that nearly all of Wachata County was present, but invalids and Rufus McPherson. I looked around for Appleseed but didn't see him anywhere.

Mr. Marshall harumphed, and clapped for attention. When things quieted down and the atmosphere was properly tense, he raised his voice like an auctioneer and called: "Now listen, everybody, in this here envelope you see in my hand" — he held a manila envelope above his head — "well, in it's the *answer* —

which nobody but God and the First National Bank knows up to now, ha, ha. And in this book" — he held up the ledger with his free hand — "I've got written down what you folks guessed. Are there any questions?" All was silence. "Fine. Now, if we could have a volunteer. . . . "

Not a living soul budged an inch: it was as if an awful shyness had overcome the crowd, and even those who were ordinarily natural-born showoffs shuffled their feet, ashamed. Then a voice, Appleseed's, hollered, "Lemme by. . . . Outa the way, please, ma'am." Trotting along behind as he pushed forward were Middy and a lanky, sleepy-eyed fellow who was evidently the fiddling brother. Appleseed was dressed the same as usual, but his face was scrubbed rosy clean, his boots polished, and his hair slicked back skin tight with Stacomb. "Did we get here in time?" he panted.

But Mr. Marshall said, "So you want to be our volunteer?"

Appleseed looked bewildered, then nodded vigorously.

"Does anybody have an objection to this young man?"

Still there was dead quiet. Mr. Marshall handed the envelope to Appleseed, who accepted it calmly. He chewed his under lip while studying it a moment before ripping his flap.

In all that congregation there was no sound except an occasional cough and the soft tinkling of Mr. R. C. Judkins' supper bell. Hamurabi was leaning against the fountain, staring up at the ceiling; Middy was gazing blankly over her brother's shoulder, and when he started to tear open the envelope, she let out a pained little gasp.

Appleseed withdrew a slip of pink paper, and holding it as though it was very fragile, muttered to himself whatever was written there. Suddenly his face paled and tears glistened in his eyes.

"Hey, speak up, boy," someone hollered.

Hamurabi stepped forward and all but snatched the slip away. He cleared his throat and commenced to read when his expression changed most comically. "Well, I'll be darned . . ." he said.

"Louder! Louder!" an angry chorus demanded.

"Buncha crooks!" yelled Mr. R. C. Judkins, who had a snootful by this time. "I smell a rat and he smells to high heaven!" Whereupon a cyclone of catcalls and whistling rent the air.

Appleseed's brother whirled round and shook his fist. "Shuddup, shuddup 'fore I bust every one of your heads together so's you got knots the size a muskmelons, hear me?"

"Citizens," cried Mayor Mawes, "citizens — I say, this is Christmas . . . I say. . . . "

And Mr. Marshall hopped up on a chair and clapped and stamped till a minimum of order was restored. It might as well be noted here that we later found out Rufus McPherson had paid Mr. R. C. Judkins to start the rumpus. Anyway, when the outbreak was quelled, who should be in possession of the slip but me . . . don't ask how.

Without thinking, I shouted, "Seventy-seven dollars and thirty-five cents." Naturally, due to the excitement, I didn't at first catch the meaning; it was just a number. Then Appleseed's brother let forth with his whooping yell, and so I understood. The name of the winner spread quickly, and the awed, murmuring whispers were like a rainstorm.

Oh, Appleseed himself was a sorry sight. He was crying as though he was mortally wounded, but when Hamurabi lifted him onto his shoulders so the crowd could get a gander, he dried his eyes with the cuffs of his sweater and began grinning. Mr. R. C. Judkins yelled, "Gyp! Lousy gyp!" but was drowned out by a deafening round of applause.

Middy grabbed my arm. "My teeth," she squealed. "Now I'm gonna get my teeth."

"Teeth?" said I, kind of dazed.

"The false kind," says she. "That's what we're gonna get us with the money — a lovely set of white false teeth."

But at that moment my sole interest was in how Appleseed had known. "Hey, tell me," I said desperately, "tell me, how in God's name did he know there was just exactly seventy-seven dollars and thirty-five cents?"

Middy gave me this *look.* "Why, I thought he told you," she said, real serious. "He counted."

"Yes, but how — how?"

"Gee, don't you even know how to count?"

"But is that all he did?"

"Well," she said, following a thoughtful pause, "he did do a little praying too." She started to dart off, then turned back and called, "Besides, he was born with a caul on his head."

And that's the nearest anybody ever came to solving the mystery. Thereafter, if you were to ask Appleseed "How come?" he would smile strangely and change the subject. Many years later he and his family moved to somewhere in Florida and were never heard from again.

But in our town his legend flourishes still; and, till his death a year ago last April, Mr. Marshall was invited each Christmas Day to tell the story of Appleseed to the Baptist Bible class. Hamurabi once typed up an account and mailed it around to various magazines. It was never printed. One editor wrote back and said that "If the little girl really turned out to be a movie star, then there might be something to your story." But that's not what happened, so why should you lie?

The Use of Force

They were new patients to me; all I had was the name, Olson. "Please come down as soon as you can; my daughter is very sick."

When I arrived, I was met by the mother, a big startled-looking woman, very clean and apologetic, who merely said, "Is this the doctor?" and let me in. "In the back," she added. "You must excuse us, doctor; we have her in the kitchen where it is warm. It is very damp here sometimes."

The child was fully dressed and sitting on her father's lap near the kitchen table. He tried to get up, but I motioned for him not to bother, took off my overcoat, and started to look things over. I could see that they were all very nervous, eyeing me up and down distrustfully. As often, in such cases, they weren't telling me more

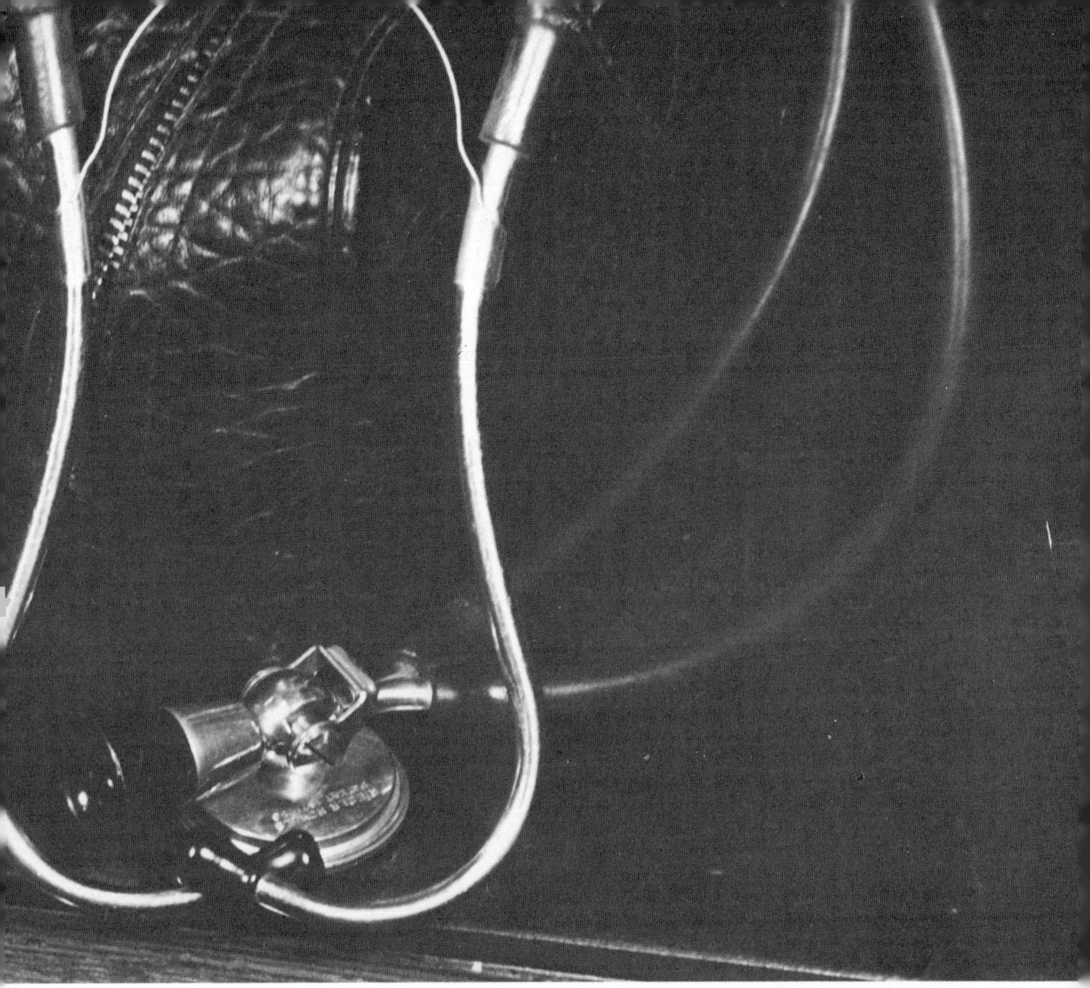

WILLIAM CARLOS WILLIAMS

than they had to; it was up to me to tell them; that's why they were spending three dollars on me.

The child was fairly eating me up with her cold, steady eyes, and no expression to her face whatever. She did not move and seemed, inwardly, quiet; an unusually attractive little thing, and as strong as a heifer[1] in appearance. But her face was flushed, she was breathing rapidly, and I realized that she had a high fever. She had magnificent blonde hair, in profusion. One of those picture children often reproduced in advertising leaflets and the photogravure sections of the Sunday papers.

[1]**heifer** (hef′ər): young cow.

"She's had a fever for three days," began the father, "and we don't know what it comes from. My wife has given her things, you know, like people do, but it don't do no good. And there's been a lot of sickness around. So we tho't you'd better look her over and tell us what is the matter."

As doctors often do, I took a trial shot at it as a point of departure. "Has she had a sore throat?"

Both parents answered me together, "No. . . . No, she says her throat don't hurt her."

"Does your throat hurt you?" added the mother to the child. But the little girl's expression didn't change, nor did she move her eyes from my face.

"Have you looked?"

"I tried to," said the mother, "but I couldn't see."

As it happens, we had been having a number of cases of diphtheria in the school to which this child went during that month, and we were all, quite apparently, thinking of that, though no one had as yet spoken of the thing.

"Well," I said, "suppose we take a look at the throat first." I smiled in my best professional manner, and asking for the child's first name, I said, "Come on, Mathilda, open your mouth and let's take a look at your throat."

Nothing doing.

"Aw, come on," I coaxed, "just open your mouth wide and let me take a look. Look," I said, opening both hands wide, "I haven't anything in my hands. Just open up and let me see."

"Such a nice man," put in the mother. "Look how kind he is to you. Come on, do what he tells you to. He won't hurt you."

At that I ground my teeth in disgust. If only they wouldn't use the word *hurt,* I might be able to get somewhere. But I did not allow myself to be hurried or disturbed, but speaking quietly and slowly I approached the child again.

As I moved my chair a little nearer, suddenly, with one catlike movement, both her hands clawed instinctively for my eyes and she almost reached them too. In fact she knocked my glasses flying and they fell, though unbroken, several feet away from me on the kitchen floor.

Both the mother and father almost turned themselves inside out in embarrassment and apology. "You bad girl," said the mother, taking her and shaking her by one arm. "Look what you've done. The nice man. . . . "

"For heaven's sake," I broke in. "Don't call me a nice man to her. I'm here to look at her throat on the chance that she might have diphtheria and possibly die of it. But that's nothing to her. Look here," I said to the child, "we're going to look at your throat. You're old enough to understand what I'm saying. Will you open it now by yourself, or shall we have to open it for you?"

Not a move. Even her expression hadn't changed. Her breaths, however, were coming faster and faster. Then the battle began. I had to do it. I had to have a throat culture[2] for her own protection. But first I told the parents that it was entirely up to them. I explained the danger but said that I would not insist on a throat examination so long as they would take the responsibility.

"If you don't do what the doctor says, you'll have to go to the hospital," the mother admonished her severely.

Oh yeah? I had to smile to myself. After all, I had already fallen in love with the savage brat; the parents were contemptible to me. In the ensuing struggle they grew more and more abject,[3] crushed, exhausted, while she surely rose to magnificent heights of insane fury of effort bred of her terror of me.

The father tried his best, and he was a big man, but the fact that she was his daughter, his shame at her behavior, and his dread of hurting her made him release her just at the critical times when I had almost achieved success, till I wanted to kill him. But his dread also that she might have diphtheria made him tell me to go on, go on, though he himself was almost fainting, while the mother moved back and forth behind us, raising and lowering her hands in an agony of apprehension.

"Put her in front of you on your lap," I ordered, "and hold both her wrists."

But as soon as he did, the child let out a scream. "Don't, you're hurting me. Let go of my hands. Let them go, I tell you." Then she shrieked terrifyingly, hysterically. "Stop it! Stop it! You're killing me!"

"Do you think she can stand it, doctor!" said the mother.

"You get out," said the husband to his wife. "Do you want her to die of diphtheria?"

"Come on, now, hold her," I said.

[2]**throat culture:** sample of bacteria to be submitted to a laboratory for examination.

[3]**abject** (ab'jekt): very humble.

Then I grasped the child's head with my left hand and tried to get the wooden tongue depressor between her teeth. She fought, with clenched teeth, desperately! But now I also had grown furious — at a child. I tried to hold myself down but I couldn't. I know how to expose a throat for inspection. And I did my best. When finally I got the wooden spatula behind the last teeth and just the point of it into the mouth cavity, she opened up for an instant, but before I could see anything, she came down again, and gripping the wooden blade between her molars, she reduced it to splinters before I could get it out again.

"Aren't you ashamed," the mother yelled at her. "Aren't you ashamed to act like that in front of the doctor?"

"Get me a smooth-handled spoon of some sort," I told the mother. "We're going through with this." The child's mouth was already bleeding. Her tongue was cut and she was screaming in wild hysterical shrieks. Perhaps I should have desisted and come back in an hour or more. No doubt it would have been better. But I have seen at least two children lying dead in bed of neglect in such cases, and feeling that I must get a diagnosis now or never, I went at it again. But the worst of it was that I too had got beyond reason. I could have torn the child apart in my own fury and enjoyed it. It was a pleasure to attack her. My face was burning with it.

The damned little brat must be protected against her own idiocy, one says to one's self at such times. Others must be protected against her. It is a social necessity. And all these things are true. But a blind fury, a feeling of adult shame, bred of a longing for muscular release, are the operatives.[4] One goes on to the end.

In a final unreasoning assault I overpowered the child's neck and jaws. I forced the heavy silver spoon back of her teeth and down her throat till she gagged. And there it was — both tonsils covered with membrane.[5] She had fought valiantly to keep me from knowing her secret. She had been hiding that sore throat for three days at least and lying to her parents in order to escape just such an outcome as this.

Now truly she was furious. She had been on the defensive before, but now she attacked. Tried to get off her father's lap and fly at me while tears of defeat blinded her eyes.

[4]**operatives:** influencing factors.
[5]**membrane:** thin layer of tissue; *here,* indicating infection.

The Little Girl and the Wolf

JAMES THURBER

One afternoon a big wolf waited in a dark forest for a little girl to come along carrying a basket of food to her grandmother. Finally a little girl did come along and she was carrying a basket of food. "Are you carrying that basket to your grandmother?" asked the wolf. The little girl said yes, she was. So the wolf asked her where her grandmother lived and the little girl told him and he disappeared into the wood.

When the little girl opened the door of her grandmother's house, she saw that there was somebody in bed with a nightcap and nightgown on. She had approached no nearer than twenty-five feet from the bed when she saw that it was not her grandmother but the wolf, for even in a nightcap a wolf does not look any more like your grandmother than the Metro-Goldwyn lion looks like Calvin Coolidge. So the little girl took an automatic out of her basket and shot the wolf dead.

Moral: It is not so easy to fool little girls nowadays as it used to be.

Adventures of Isabel

OGDEN NASH

Isabel met an enormous bear,
Isabel, Isabel, didn't care;
The bear was hungry, the bear was ravenous,
The bear's big mouth was cruel and cavernous.
The bear said, Isabel, glad to meet you,
How do, Isabel, now I'll eat you!
Isabel, Isabel, didn't worry,
Isabel didn't scream or scurry.
She washed her hands and she straightened her hair up,
Then Isabel quietly ate the bear up.

Once in a night as black as pitch
Isabel met a wicked old witch.
The witch's face was cross and wrinkled,
The witch's gums with teeth were sprinkled.
Ho, ho, Isabel! the old witch crowed,
I'll turn you into an ugly toad!
Isabel, Isabel, didn't worry,
Isabel didn't scream or scurry.
She showed no rage and she showed no rancor,[1]
But she turned the witch into milk and drank her.

Isabel met a hideous giant,
Isabel continued self-reliant.
The giant was hairy, the giant was horrid,
He had one eye in the middle of his forehead.

[1]**rancor** (rang′kər): bitter resentment.

Good morning, Isabel, the giant said,
I'll grind your bones to make my bread.
Isabel, Isabel, didn't worry,
Isabel didn't scream or scurry.
She nibbled the zwieback[2] that she always fed off,
And when it was gone, she cut the giant's head off.

Isabel met a troublesome doctor,
He punched and poked till he really shocked her.
The doctor's talk was of coughs and chills
And the doctor's satchel bulged with pills.
The doctor said unto Isabel,
Swallow this, it will make you well.
Isabel, Isabel, didn't worry,
Isabel didn't scream or scurry.
She took those pills from the pill-concocter,
And Isabel calmly cured the doctor.

[2]**zwieback** (swī'bak): sweet toasted bread.

"Tell it like it never really was,
man,
and maybe we can see it
like it is."

Edward Lueders

Tom Edison's Shaggy Dog

KURT VONNEGUT, JR.

Two old men sat on a park bench one morning in the sunshine of Tampa, Florida — one trying doggedly to read a book he was plainly enjoying, while the other, Harold K. Bullard, told him the story of his life in the full, round, head tones of a public-address system. At their feet lay Bullard's Labrador retriever, who further tormented the aged listener by probing his ankles with a large, wet nose.

Bullard, who had been, before he retired, successful in many fields, enjoyed reviewing his important past. But he faced the problem that complicates the lives of cannibals — namely: that a single victim cannot be used over and over. Anyone who had passed the time of day with him and his dog refused to share a bench with them again.

Shaggy Dog: A "shaggy dog" story is a long, rambling tale, sometimes about a talking animal.

So Bullard and his dog set out through the park each day in quest of new faces. They had had good luck this morning, for they had found this stranger right away, clearly a new arrival in Florida, still buttoned up tight in heavy serge, stiff collar, and necktie, and with nothing better to do than read.

"Yes," said Bullard, rounding out the first hour of his lecture, "made and lost five fortunes in my time."

"So you said," said the stranger, whose name Bullard had neglected to ask. "Easy, boy. No, no, no, boy," he said to the dog, who was growing more aggressive toward his ankles.

"Oh? Already told you that, did I?" said Bullard.

"Twice."

"Two in real estate, one in scrap iron, and one in oil and one in trucking."

"So you said."

"I did? Yes, guess I did. Two in real estate, one in scrap iron, one in oil, and one in trucking. Wouldn't take back a day of it."

"No, I suppose not," said the stranger. "Pardon me, but do you suppose you could move your dog somewhere else? He keeps —"

"Him?" said Bullard, heartily. "Friendliest dog in the world. Don't need to be afraid of him."

"I'm not afraid of him. It's just that he drives me crazy, sniffing at my ankles."

"Plastic," said Bullard, chuckling.

"What?"

"Plastic. Must be something plastic on your garters. By golly, I'll bet it's those little buttons. Sure as we're sitting here, those buttons must be plastic. That dog is nuts about plastic. Don't know why that is, but he'll sniff it out and find it if there's a speck around. Must be a deficiency in his diet, though by gosh, he eats better than I do. Once he chewed up a whole plastic humidor.[1] Can you beat it? *That's* the business I'd go into now, by glory, if the pill rollers hadn't told me to let up, to give the old ticker a rest."

"You could tie the dog to that tree over there," said the stranger.

"I get so darn' sore at all the youngsters these days!" said Bullard. "All of 'em mooning around about no frontiers any more.

[1]**humidor** (hū'mə·dôr): box or jar for keeping tobacco moist.

There never have been so many frontiers as there are today. You know what Horace Greeley[2] would say today?"

"His nose is wet," said the stranger, and he pulled his ankles away, but the dog humped forward in patient pursuit. "Stop it, boy!"

"His wet nose shows he's healthy," said Bullard. "'Go plastic, young man!' That's what Greeley'd say. 'Go atom, young man!'"

The dog had definitely located the plastic buttons on the stranger's garters and was cocking his head one way and another, thinking out ways of bringing his teeth to bear on those delicacies.

"Scat!" said the stranger.

"'Go electronic, young man!'" said Bullard. "Don't talk to me about no opportunity any more. Opportunity's knocking down every door in the country, trying to get in. When I was young, a man had to go out and find opportunity and drag it home by the ears. Nowadays —"

"Sorry," said the stranger, evenly. He slammed his book shut, stood, and jerked his ankle away from the dog. "I've got to be on my way. So good day, sir."

He stalked across the park, found another bench, sat down with a sigh, and began to read. His respiration had just returned to normal when he felt the wet sponge of the dog's nose on his ankles again.

[2]**Horace Greeley:** American journalist and political leader famous for saying "Go West, young man, go West."

"Oh—it's you!" said Bullard, sitting down beside him. "He was tracking you. He was on the scent of something, and I just let him have his head. What'd I tell you about plastic?" He looked about contentedly. "Don't blame you for moving on. It was stuffy back there. No shade to speak of and not a sign of a breeze."

"Would the dog go away if I bought him a humidor?" said the stranger.

"Pretty good joke, pretty good joke," said Bullard, amiably. Suddenly he clapped the stranger on his knee. "Sa-ay, you aren't in plastics, are you? Here I've been blowing off about plastics, and for all I know that's your line."

"My line?" said the stranger crisply, laying down his book. "Sorry — I've never had a line. I've been a drifter since the age of nine, since Edison set up his laboratory next to my home and showed me the intelligence analyzer."

"Edison?" said Bullard. "Thomas Edison, the inventor?"

"If you want to call him that, go ahead," said the stranger.

"If I *want* to call him that?"—Bullard guffawed—"I guess I just will! Father of the light bulb and I don't know what all."

"If you want to think he invented the light bulb, go ahead. No harm in it." The stranger resumed his reading.

"Say, what is this?" said Bullard, suspiciously. "You pulling my leg? What's this about an intelligence analyzer? I never heard of that."

"Of course you haven't," said the stranger. "Mr. Edison and I promised to keep it a secret. I've never told anyone. Mr. Edison broke his promise and told Henry Ford, but Ford made him promise not to tell anybody else — for the good of humanity."

Bullard was entranced. "Uh, this intelligence analyzer," he said, "it analyzed intelligence, did it?"

"It was an electric butter churn," said the stranger.

"Seriously now," Bullard coaxed.

"Maybe it *would* be better to talk it over with someone," said the stranger. "It's a terrible thing to keep bottled up inside me, year in and year out. But how can I be sure that it won't go any further?"

"My word as a gentleman," Bullard assured him.

"I don't suppose I could find a stronger guarantee than that, could I?" said the stranger, judiciously.[3]

[3]**judiciously** (jü·dish'əs·lē): wisely; sensibly.

"There is no stronger guarantee," said Bullard proudly. "Cross my heart and hope to die!"

"Very well." The stranger leaned back and closed his eyes, seeming to travel backward through time. He was silent for a full minute, during which Bullard watched with respect.

"It was back in the fall of eighteen seventy-nine," said the stranger at last, softly. "Back in the village of Menlo Park, New Jersey. I was a boy of nine. A young man we all thought was a wizard had set up a laboratory next door to my home, and there were flashes and crashes inside, and all sorts of scary goings-on. The neighborhood children were warned to keep away, not to make any noise that would bother the wizard.

"I didn't get to know Edison right off, but his dog Sparky and I got to be steady pals. A dog a whole lot like yours, Sparky was, and we used to wrestle all over the neighborhood. Yes, sir, your dog is the image of Sparky."

"Is that so?" said Bullard, flattered.

"Gospel," replied the stranger. "Well, one day Sparky and I were wrestling around, and we wrestled right up to the door of Edison's laboratory. The next thing I knew, Sparky had pushed me in through the door, and bam! I was sitting on the laboratory floor, looking up at Mr. Edison himself."

"Bet he was sore," said Bullard, delighted.

"You can bet I was scared," said the stranger. "I thought I was face to face with Satan himself. Edison had wires hooked to his ears and running down to a little black box in his lap! I started to scoot, but he caught me by my collar and made me sit down.

"'Boy,' said Edison, 'it's always darkest before the dawn. I want you to remember that.'

"'Yes, sir,' I said.

"'For over a year, my boy,' Edison said to me, 'I've been trying to find a filament[4] that will last in an incandescent[5] lamp. Hair, string, splinters — nothing works. So while I was trying to think of something else to try, I started tinkering with another idea of mine, just letting off steam. I put this together,' he said, showing me the little black box. 'I thought maybe intelligence was just a certain kind of electricity, so I made this intelligence analyzer here. It works!

[4]**filament:** threadlike conductor of electricity.
[5]**incandescent** (in·kən·des'nt): glowing from great heat.

You're the first one to know about it, my boy. But I don't know why you shouldn't be. It will be your generation that will grow up in the glorious new era when people will be as easily graded as oranges.'"

"I don't believe it!" said Bullard.

"May I be struck by lightning this very instant!" said the stranger. "And it did work, too. Edison had tried out the analyzer on the men in his shop, without telling them what he was up to. The smarter a man was, by gosh, the farther the needle on the indicator in the little black box swung to the right. I let him try it on me, and the needle just lay where it was and trembled. But dumb as I was, then is when I made my one and only contribution to the world. As I say, I haven't lifted a finger since."

"Whadja do?" said Bullard, eagerly.

"I said, 'Mr. Edison, sir, let's try it on the dog.' And I wish you could have seen the show that dog put on when I said it! Old

Sparky barked and howled and scratched to get out. When he saw we meant business, that he wasn't going to get out, he made a beeline right for the intelligence analyzer and knocked it out of Edison's hands. But we cornered him, and Edison held him down while I touched the wires to his ears. And would you believe it, that needle sailed clear across the dial, way past a little red pencil marker on the dial face!"

"The dog busted it," said Bullard.

"'Mr. Edison, sir,' I said, 'what's the red mark mean?'

"'My boy,' said Edison, 'it means that the instrument is broken, because that red mark is me.'"

"I'll say it was broken," said Bullard.

The stranger said gravely, "But it wasn't broken. No, sir. Edison checked the whole thing, and it was in apple-pie order. When Edison told me that, it was then that Sparky, crazy to get out, gave himself away."

"How?" said Bullard, suspiciously.

"We really had him locked in, see? There were three locks on the door — a hook and eye, a bolt, and a regular knob and latch. That dog stood up, unhooked the hook, pushed the bolt back, and had the knob in his teeth when Edison stopped him."

"No!" said Bullard.

"Yes!" said the stranger, his eyes shining. "And then is when Edison showed me what a great scientist he was. He was willing to face the truth, no matter how unpleasant it might be.

"'So!' said Edison to Sparky. 'Man's best friend, huh? Dumb animal, huh?'

"That Sparky was a caution. He pretended not to hear. He scratched himself and bit fleas and went around growling at ratholes — anything to get out of looking Edison in the eye.

"'Pretty soft, isn't it, Sparky?' said Edison. 'Let somebody else worry about getting food, building shelters, and keeping warm, while you sleep in front of a fire or go chasing after the girls or raise hell with the boys. No mortgages, no politics, no war, no work, no worry. Just wag the old tail or lick a hand, and you're all taken care of.'

"'Mr. Edison,' I said, 'do you mean to tell me that dogs are smarter than people?'

"'Smarter?' said Edison. 'I'll tell the world! And what have I been doing for the past year? Slaving to work out a light bulb so dogs can play at night!'

"'Look, Mr. Edison,' said Sparky, 'why not—'"

"Hold on!" roared Bullard.

"Silence!" shouted the stranger, triumphantly. "'Look, Mr. Edison,' said Sparky, 'why not keep quiet about this? It's been working out to everybody's satisfaction for hundreds of thousands of years. Let sleeping dogs lie. You forget all about it, destroy the intelligence analyzer, and I'll tell you what to use for a lamp filament.'"

"Hogwash!" said Bullard, his face purple.

The stranger stood. "You have my solemn word as a gentleman. That dog rewarded *me* for my silence with a stock-market tip that made me independently wealthy for the rest of my days. And the last words that Sparky ever spoke were to Thomas Edison. 'Try a piece of carbonized cotton thread,' he said. Later, he was torn to bits by a pack of dogs that had gathered outside the door, listening."

The stranger removed his garters and handed them to Bullard's dog. "A small token of esteem, sir, for an ancestor of yours who talked himself to death. Good day." He tucked his book under his arm and walked away.

Your Poem, Man...

EDWARD LUEDERS

unless there's one thing seen
suddenly against another — a parsnip
sprouting for a President, or
hailstones melting in an ashtray —
nothing really happens. It takes
surprise and wild connections,
doesn't it? A walrus chewing
on a ballpoint pen. Two blue tail-
lights on Tyrannosaurus Rex.[1] Green
cheese teeth. Maybe what we wanted
least. Or most. Some unexpected
pleats. Words that never knew
each other till right now. Plug us
into the wrong socket and see
what blows—or what lights up.
Try
        untried
                circuitry,
new
        fuses.
Tell it like it never really was,
man,
and maybe we can see it
like it is.

[1]**Tyrannosaurus Rex** (tə·ran′ə·sôr·əs rex′): type of dinosaur.

The Talking Skull

NUPE FOLK TALE

A hunter goes into the bush. He finds an old human skull. The hunter says: "What brought you here?" The skull answers: "Talking brought me here." The hunter runs off. He runs to the king. He tells the king: "I found a dry human skull in the bush. It asks you how its father and mother are."

The king says: "Never since my mother bore me have I heard that a dead skull can speak." The king summons the Alkali, the Saba, and the Degi and asks them if they have ever heard the like. None of the wise men has heard the like, and they decide to send a guard out with the hunter into the bush to find out if his story is true, and if so, to learn the reason for it. The guard accompany the hunter into the bush with the order to kill him on the spot should he have lied. The guard and the hunter come to the skull. The hunter addresses the skull: "Skull, speak." The skull is silent. The hunter asks as before: "What brought you here?" The skull does not answer. The whole day long the hunter begs the skull to speak, but it does not answer. In the evening the guard tell the hunter to make the skull speak, and when he cannot, they kill him in accordance with the king's command. When the guard are gone, the skull opens its jaws and asks the dead hunter's head: "What brought you here?" The dead hunter's head replies: "Talking brought me here!"

Nupe (nü'pā): West African people of the Niger River Valley.

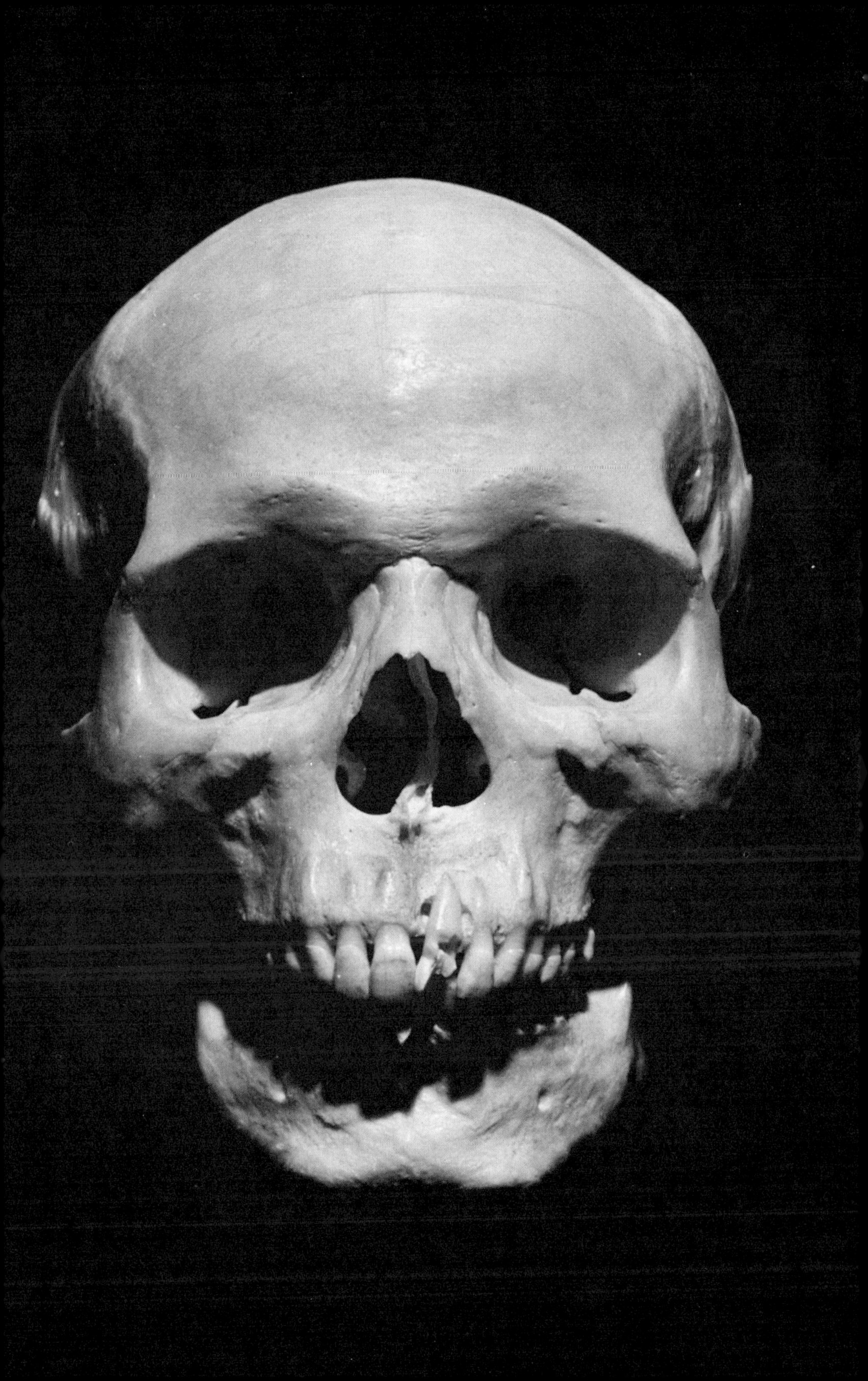

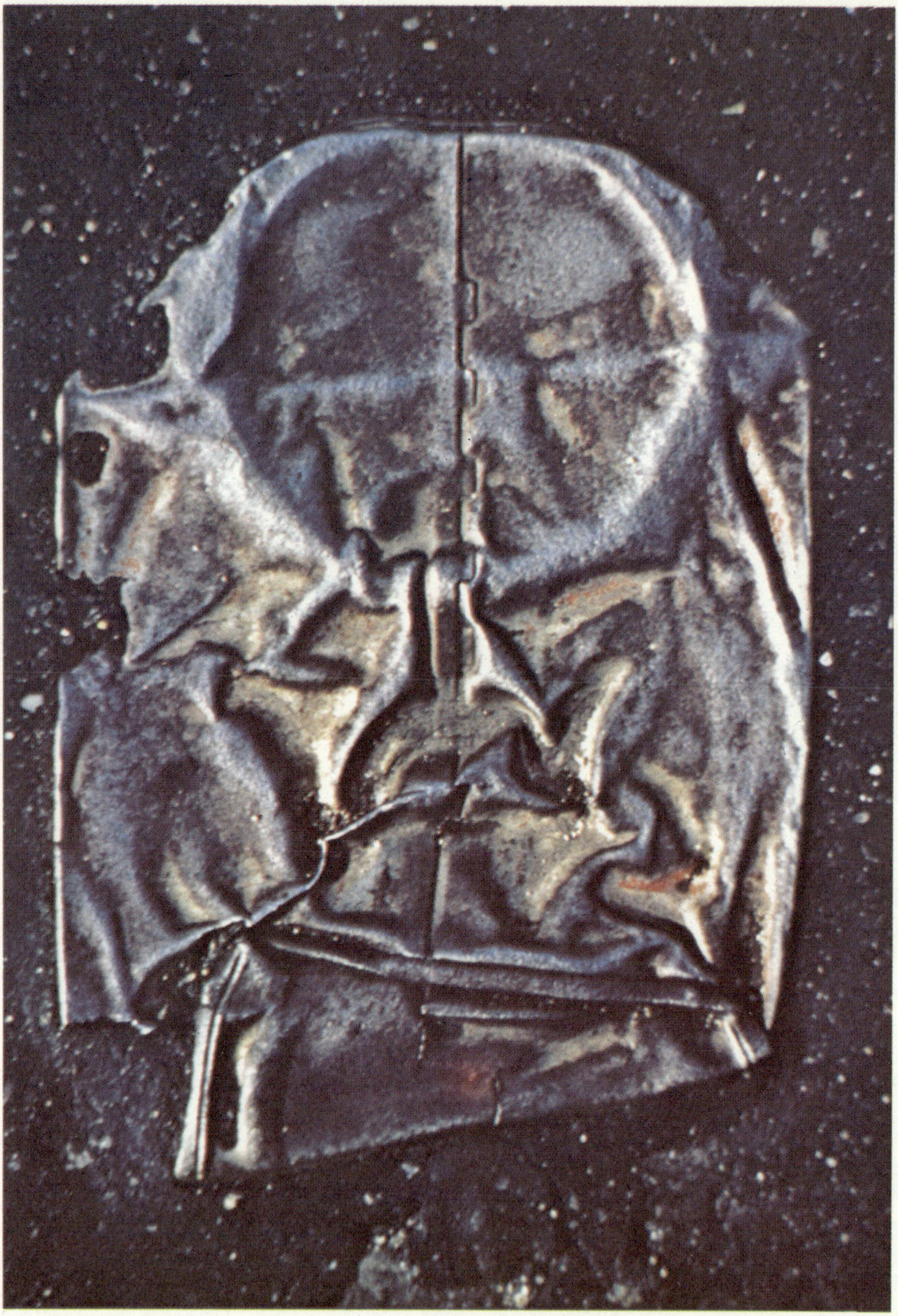

The strength of a mother is expressed
in the giving of life.
Sometimes that life may be her own.

All the Years of Her Life

MORLEY CALLAGHAN

They were closing the drugstore, and Alfred Higgins, who had just taken off his white jacket, was putting on his coat and getting ready to go home. The little gray-haired man, Sam Carr, who owned the drugstore, was bending down behind the cash register, and when Alfred Higgins passed him, he looked up and said softly, "Just a moment, Alfred. One moment before you go."

The soft, confident, quiet way in which Sam Carr spoke made Alfred start to button his coat nervously. He felt sure his face was white. Sam Carr usually said, "Good night," brusquely, without looking up. In the six months he had been working in the drugstore Alfred had never heard his employer speak softly like that. His heart began to beat so loud it was hard for him to get his breath. "What is it, Mr. Carr?" he asked.

"Maybe you'd be good enough to take a few things out of your pocket and leave them here before you go," Sam Carr said.

"What things? What are you talking about?"

"You've got a compact and a lipstick and at least two tubes of toothpaste in your pockets, Alfred."

"What do you mean? Do you think I'm crazy?" Alfred blustered. His face got red, and he knew he looked fierce with indignation. But Sam Carr, standing by the door with his blue eyes shining bright behind his glasses and his lips moving underneath his gray mustache, only nodded his head a few times, and then Alfred grew very frightened and he didn't know what to say. Slowly he raised his hand and dipped it into his pocket, and with his eyes never meeting Sam Carr's eyes, he took out a blue compact and two tubes of toothpaste and a lipstick, and he laid them one by one on the counter.

"Petty thieving, eh, Alfred?" Sam Carr said. "And maybe you'd be good enough to tell me how long this has been going on."

"This is the first time I ever took anything."

"So now you think you'll tell me a lie, eh? What kind of a sap do I look like, huh? I don't know what goes on in my own store, eh? I tell you you've been doing this pretty steady," Sam Carr said as he went over and stood behind the cash register.

Ever since Alfred had left school, he had been getting into trouble wherever he worked. He lived at home with his mother and his father, who was a printer. His two older brothers were married, and his sister had got married last year, and it would have been all right for his parents now if Alfred had only been able to keep a job.

While Sam Carr smiled and stroked the side of his face very delicately with the tips of his fingers, Alfred began to feel that familiar terror growing in him that had been in him every time he had got into such trouble.

"I liked you," Sam Carr was saying. "I liked you and would have trusted you, and now look what I got to do." While Alfred watched with his alert, frightened blue eyes, Sam Carr drummed with his fingers on the counter. "I don't like to call a cop in point-blank," he was saying as he looked very worried. "You're a fool, and maybe I should call your father and tell him you're a fool. Maybe I should let them know I'm going to have you locked up."

"My father's not at home. He's a printer. He works nights," Alfred said.

"Who's at home?"

"My mother, I guess."

"Then we'll see what she says." Sam Carr went to the phone and dialed the number. Alfred was not so much ashamed, but there was that deep fright growing in him, and he blurted out arrogantly, like a strong, full-grown man, "Just a minute. You don't need to draw anybody else in. You don't need to tell her." He wanted to sound like a swaggering, big guy who could look after himself, yet the old, childish hope was in him, the longing that someone at home would come and help him. "Yeah, that's right, he's in trouble," Mr. Carr was saying. "Yeah, your boy works for me. You'd better come down in a hurry." And when he was finished, Mr. Carr went over to the door and looked out at the street and watched the people passing in the late summer night. "I'll keep my eye out for a cop" was all he said.

Alfred knew how his mother would come rushing in; she would rush in with her eyes blazing, or maybe she would be crying, and she would push him away when he tried to talk to her, and make him feel her dreadful contempt; yet he longed that she might come before Mr. Carr saw the cop on the beat passing the door.

While they waited — and it seemed a long time — they did not speak, and when at last they heard someone tapping on the closed door, Mr. Carr, turning the latch, said crisply, "Come in, Mrs. Higgins." He looked hard-faced and stern.

Mrs. Higgins must have been going to bed when he telephoned, for her hair was tucked in loosely under her hat, and her hand at her throat held her light coat tight across her chest so her dress would not show. She came in, large and plump, with a little smile on her friendly face. Most of the store lights had been turned out, and at first she did not see Alfred, who was standing in the shadow at the end of the counter. Yet as soon as she saw him, she did not look as Alfred thought she would look: she smiled, her blue eyes never wavered, and with a calmness and dignity that made them forget that her clothes seemed to have been thrown on her, she put out her hand to Mr. Carr and said politely, "I'm Mrs. Higgins. I'm Alfred's mother."

Mr. Carr was a bit embarrassed by her lack of terror and her simplicity, and he hardly knew what to say to her, so she asked, "Is Alfred in trouble?"

"He is. He's been taking things from the store. I caught him red-handed. Little things like compacts and toothpaste and lipsticks. Stuff he can sell easily," the proprietor said.

As she listened, Mrs. Higgins looked at Alfred sometimes and nodded her head sadly, and when Sam Carr had finished, she said gravely, "Is it so, Alfred?"

"Yes."

"Why have you been doing it?"

"I been spending money, I guess."

"On what?"

"Going around with the guys, I guess," Alfred said.

Mrs. Higgins put out her hand and touched Sam Carr's arm with an understanding gentleness, and speaking as though afraid of disturbing him, she said, "If you would only listen to me before doing anything." Her simple earnestness made her shy; her humility made her falter and look away, but in a moment she was smiling gravely again, and she said with a kind of patient dignity, "What did you intend to do, Mr. Carr?"

"I was going to get a cop. That's what I ought to do."

"Yes, I suppose so. It's not for me to say, because he's my son. Yet I sometimes think a little good advice is the best thing for a boy when he's at a certain period in his life," she said.

Alfred couldn't understand his mother's quiet composure, for if they had been at home and someone had suggested that he was going to be arrested, he knew she would be in a rage and would cry out against him. Yet now she was standing there with that gentle, pleading smile on her face, saying, "I wonder if you don't think it would be better just to let him come home with me. He looks a big fellow, doesn't he? It takes some of them a long time to get any sense," and they both stared at Alfred, who shifted away with a bit of light shining for a moment on his thin face and the tiny pimples over his cheekbone.

But even while he was turning away uneasily, Alfred was realizing that Mr. Carr had become aware that his mother was really a fine woman; he knew that Sam Carr was puzzled by his mother, as if he had expected her to come in and plead with him tearfully, and instead he was being made feel a bit ashamed by her vast tolerance. While there was only the sound of the mother's soft, assured voice in the store, Mr. Carr began to nod his head encouragingly at her. Without being alarmed, while being just large and still and simple and hopeful, she was becoming dominant there in the dimly lit store. "Of course, I don't want to be harsh," Mr. Carr was saying. "I'll tell you what I'll do. I'll just fire him and let it go at that. How's that?" and he got up and shook hands with Mrs. Higgins, bowing low to her in deep respect.

There was such warmth and gratitude in the way she said, "I'll never forget your kindness," that Mr. Carr began to feel warm and genial himself.

"Sorry we had to meet this way," he said. "But I'm glad I got in touch with you. Just wanted to do the right thing, that's all," he said.

"It's better to meet like this than never, isn't it?" she said. Suddenly they clasped hands as if they liked each other, as if they had known each other a long time. "Good night, sir," she said.

"Good night, Mrs. Higgins. I'm truly sorry," he said.

The mother and son walked along the street together, and the mother was taking a long, firm stride as she looked ahead with her stern face full of worry. Alfred was afraid to speak to her; he was afraid of the silence that was between them, so he only looked ahead too, for the excitement and relief was still pretty strong in him; but in a little while, going along like that in silence made him terribly aware of the strength and the sternness in her; he began to wonder what she was thinking of as she stared ahead so grimly; she seemed to have forgotten that he walked beside her; so when they were passing under the Sixth Avenue elevated and the rumble of the train seemed to break the silence, he said in his old, blustering way, "Thank God it turned out like that. I certainly won't get in a jam like that again."

"Be quiet. Don't speak to me. You've disgraced me again and again," she said bitterly.

"That's the last time. That's all I'm saying."

"Have the decency to be quiet," she snapped. They kept on their way, looking straight ahead.

When they were at home and his mother took off her coat, Alfred saw that she was really only half dressed, and she made him feel afraid again when she said, without even looking at him, "You're a bad lot. God forgive you. It's one thing after another and always has been. Why do you stand there stupidly? Go to bed, why don't you?" When he was going, she said, "I'm going to make myself a cup of tea. Mind, now, not a word about tonight to your father."

While Alfred was undressing in his bedroom, he heard his mother moving around the kitchen. She filled the kettle and put it on the stove. She moved a chair. And as he listened, there was no shame in him, just wonder and a kind of admiration of her strength and repose. He could still see Sam Carr nodding his head encouragingly to her; he could hear her talking simply and earnestly, and as he sat on his bed, he felt a pride in her strength. "She certainly was smooth," he thought. "Gee, I'd like to tell her she sounded swell."

And at last he got up and went along to the kitchen, and when he was at the door, he saw his mother pouring herself a cup of tea. He watched and he didn't move. Her face, as she sat there, was a frightened, broken face utterly unlike the face of the woman who had been so assured a little while ago in the drugstore. When she reached out and lifted the kettle to pour hot water in her cup, her hand trembled and the water splashed on the stove. Leaning back in the chair, she sighed and lifted the cup to her lips, and her lips were groping loosely as if they would never reach the cup. She swallowed the hot tea eagerly, and then she straightened up in relief, though her hand holding the cup still trembled. She looked very old.

It seemed to Alfred that this was the way it had been every time he had been in trouble before, that this trembling had really been in her as she hurried out half dressed to the drugstore. He understood why she had sat alone in the kitchen the night his young sister had kept repeating doggedly that she was getting married. Now he felt all that his mother had been thinking of as they walked along the street together a little while ago. He watched his mother, and he never spoke, but at that moment his youth seemed to be over; he knew all the years of her life by the way her hand trembled as she raised the cup to her lips. It seemed to him that this was the first time he had ever looked upon his mother.

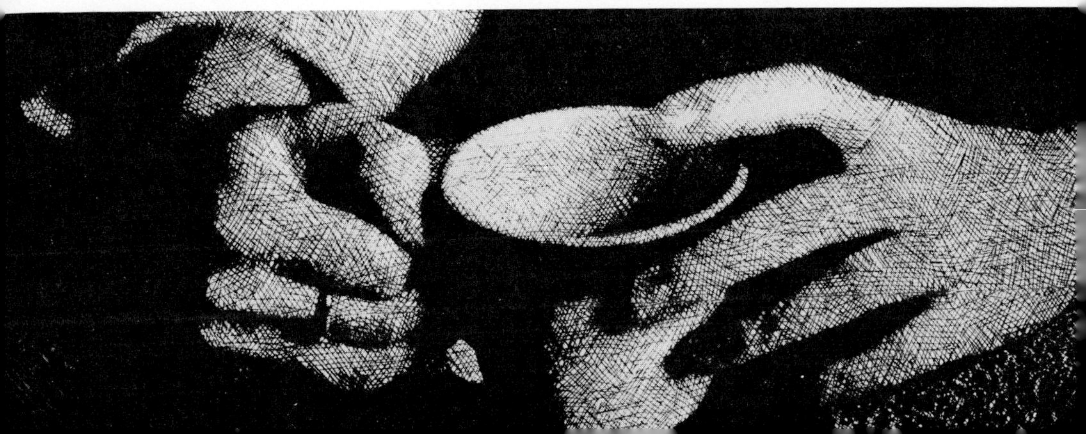

These episodes from the novel When the Legends Die *tell of Ute Indians in Colorado some sixty years ago. When George Black Bull kills Frank No Deer in a fight, he flees to the mountains with his wife Bessie and their five-year-old son Thomas. During their third winter there, George dies in an avalanche. In the spring, Bessie goes to trade in the town of Pagosa and learns that the authorities have ruled that George killed Frank in self-defense. The following selection begins a year later.*

Bessie Black Bull

HAL BORLAND

They spent that winter as before. Spring came again, and one day when he went to catch fish at a pool in the stream, he met the she-bear again. He was not afraid of her. He said to her, "You gave me my name. I am Bear Brother and we are friends." She listened as he sang a song that came to him, a song of friendship with that bear. Then she went away and he followed her and saw her uproot an old stump and catch three chipmunks and lick up the swarming ants. Then he went to the stream and caught fish, and he left half of his fish for the bear.

He told his mother of this and she said, "It is good to have a friend." She told him how their people had been friends of everything in the mountains in the old days — the bears, the deer, the mountain lions, the jays, the ravens. She told him these things. Then she was silent. He waited for her to say more, but she could not tell him the things that were in her mind. Ever since they had gone to Pagosa, those things had been in her mind, the things Jim Thatcher had said about her man. She had been so startled, so afraid, that she hadn't really heard what he said except that she should tell her man that the matter of killing Frank No Deer was all settled, over with. She wanted to hear these things again. This time she would listen. She would not be afraid this time.

Then it was summer. Soon the serviceberries would be ripe for picking, and they should smoke fish when the berries were ripe. She said to the boy, "I must go on a journey."

He said, "When you say, we shall go."

She said, "I must go alone. You will stay here until I come back." She still worried a little that they might want the boy. If they took her, she could manage to get away, somehow. But the boy might not know how to get away. She said, "You must stay here."

That is the way it was. She went to Pagosa again, and the boy stayed there on Bald Mountain.

She went to Pagosa. She took two baskets to trade at Jim Thatcher's store. She did not need an ax or a knife, but she took the baskets to trade so that she would have a reason to listen when Jim Thatcher talked to her, asked questions.

Jim Thatcher liked the baskets, as before. He asked her what she wanted to trade them for, and she chose calico for a skirt, and cloth for a blouse, and a blue coat with brass buttons for the boy. It was more than he would give her for the baskets. She knew this. She wanted him to talk.

He said, "This is too much, Bessie."

She pretended she did not understand. She took one basket from the counter.

He said, "*This* is too much," and he put a hand on what she had chosen.

They haggled. Finally he said, "You are a smart woman, Bessie. But you didn't tell your man what I said to tell him. He hasn't been to town."

"There is this thing about Frank No Deer," she said.

"I told you that was all settled."

She shrugged.

"You didn't believe me?"

She smiled, deprecating.[1]

"Well, it is. All settled." He made the gesture, cut-off, finished. Then he asked, "Where is George?"

She made the gesture he had just made. Cut-off, finished. It also meant dead.

"That's too bad. What happened?"

"The slide, the snowslide." She unconsciously fell into English. "He was hunting. The slide came. He is dead, two winters gone."

He watched her, knew she was telling the truth. "I'm sorry, Bessie. And the boy?"

She smiled. Then she held out a hand, held it as high as her own head. She put her hand on the blue jacket. "For him," she said.

Jim Thatcher went to a rack and got a denim jacket the same size. Something more practical. But she shook her head. She wanted the serge jacket with the brass buttons. He looked at the baskets again, figured that they would just about cover her purchases, and gave in. He was tempted to throw in a bag of candy. But she hadn't asked for it. He let it pass.

She put the purchases in her pack and left the store. She went down the street, and at the corner by the bank, with its big windows and brick front, she met Blue Elk.[2]

Blue Elk saw her and stopped, his mouth shaped in surprise.

[1]**deprecating** (dep′rə·kāt·ing): disapproving; protesting.

[2]**Blue Elk:** Indian who for a price would negotiate with white men in behalf of other Indians.

He said, "I have waited a long time to see you."

She said, "I am here. You see me." Her voice was cold.

"We must talk," Blue Elk said.

She said, "I will listen."

They went across the street and down toward the bridge where there was a vacant lot. There she sat down and put the pack beside her. Blue Elk sat awkwardly. He was used to sitting on a chair. He took off his derby hat and she saw streaks of gray in his hair, though his braids were still black. His cheeks hung down like the cheeks of an old dog and his belly was even fatter than she remembered.

He said, "Where is your man? And the boy?"

She shrugged.

"He can come back," he said. "I settled that thing. I took care of it. He can come back and work in the sawmill again."

"No," she said.

He glanced at her pack, then at her clothes. "You are living in the old way," he said.

She didn't answer.

"Where?" he asked.

Still she didn't answer.

He sat silent a long moment, then said, "I settled this thing for your man."

She said, "It is settled."

"I did it. I said I would do it for the cost of two horses. You owe me the cost of two horses."

"I told you I did not have the cost of one goat. I have no horses. I do not even have one goat." And she made the cut-off sign. She said with the sign that it was finished.

He said, "They did not follow you when you left."

She did not answer.

"They did not follow you to find him," he insisted.

"They did not find him," she said.

"It is as I said. I told them he killed Frank No Deer because Frank tried to kill him. That is the way it was settled."

"Frank No Deer stole his money. Three times he stole it. That was enough."

Blue Elk shook his head. "That was not enough. To be a thief is not enough."

"This is a strange thing you say," she told him angrily. "You say it is not bad to steal my man's money three times, but it is bad to kill Frank No Deer for that."

"I say it is the law. That is the way it is."

"I do not like this law."

"I settled this. You owe to me for settling this."

"My man will pay you."

"Where is your man?"

She looked at him and she made the cut-off sign.

"Dead?" he asked, and he unconsciously put his hand to his mouth for saying it.

She nodded.

He shook his head, sat silent for a time. Then he said, "The boy?"

She did not answer.

"The boy should be in the school at Ignacio."

"No."

"That is where he should be."

"No!"

He stretched his legs, stiff from sitting on the ground. He reached for her pack. She held it from him. He grabbed it, pulled it open. The cloth for the skirt and the cloth for the blouse and the blue coat with brass buttons fell out. He reached for them. She caught up the blue coat and held it away from him. He took the cloth. He said, "You owe me for settling this thing for your man. I will keep this cloth. Give me the coat."

"No." She tried to take the empty pack from him.

He was on his knees, holding to the pack, holding the cloth in his other hand. She had the coat. She turned and ran from him, to the street and up the hill away from the bridge. There she ran down the street and out of Pagosa. She went a little way along the road, then left it and went up the hillside, where she knew her way. She started back to Bald Mountain, with no pack, no food. But with the blue coat with brass buttons. And with the knowledge that it was settled, this thing about her man and Frank No Deer. It was three days back to the lodge, but there were berries to eat, and there were roots.

It was another summer before she went to Pagosa again. She went alone, as before. This time she took four baskets to trade.

She was no longer afraid of the sheriff or the sawmill man, but when she came to two Indian women gathering firewood on the hill near Pagosa, she stopped to talk to them. She did not know these women. They seemed not to know her, for they looked at her with curious eyes and smiled at her clothes and looked at her pack. She made the sign of greeting and she asked if Blue Elk was in Pagosa. The one with a purple blouse pointed with her chin toward Bessie's pack and said to the other, "This one has a gift for Blue Elk." The other one laughed. She said, "No. This one thinks Blue Elk has a gift for her. That is why she brings the big pack, to carry it in." They both laughed.

Bessie said, "I have come to trade with Jim Thatcher. Blue Elk stole skirt cloth and blouse cloth from me last summer."

They said, "Oh-oh." Then the one with a purple blouse said, "Fat Belly is not here. He is at the reservation."

Bessie said, "This is true?"

They both said, "This is true. He is not here to steal from you. We know that one."

She helped them gather wood for a little time, to pay them for this. Then she went on to Pagosa.

She went to Jim Thatcher's store and she traded the four baskets for a red blanket. Not the Navaho kind, which were black and white and gray as well as red, but the white man's kind, which was all red. When she had finished trading and put the red blanket in her pack, Jim Thatcher said, "Watch out for Blue Elk this time, Bessie."

"That one!" She said it like a curse.

"He tried to sell that skirt cloth back to me last summer. I wouldn't take it and he made quite a fuss. What happened?"

"He stole that cloth from me."

"Where?"

"Here. In Pagosa."

"Why didn't *you* make a fuss?"

She shrugged. "He said I owed him the cost of two horses."

"What for?"

"For settling that thing about my man."

"Why, the old scoundrel! He wasn't even a witness at the inquest. Well, he's out of town today."

So she went back to Bald Mountain. She gave the red blanket to the boy for his bed. She said she wanted him to sleep warm in the winter. She did not say, even to herself, that red is the rainbow color for protection. The boy knew this. She had told him this when she told him the old tales.

The boy had begun to fill out, to have the stocky frame typical of the Utes. He could drive an arrow all the way through a deer, and he could carry the hindquarters of a buck on his shoulders. He had braids almost as long as those of his mother and he kept them wrapped with red wool at their ends. He wore a breechclout and moccasins in the summer and in the winter he wore the winter leggings and the winter shirt. Now he had the red blanket.

That was a bad winter. There was much snow and the deer moved to the lower valleys. Even the rabbits were scarce that winter. Their dried fish and cured meat were gone before the daylight began to lengthen, even though she ate little, saving the meat for the boy. "You must be strong," she said. "When these storms end, you must be strong to go to the lower valleys and get meat."

BOBOLINK
12-OZ
7-UP
ENRICHED
FLOUR
BLEACHED

At last the storms eased. He prepared to get meat. She said she would go with him and he said she should not do that. She remembered what had happened to her man at such a time. She did not say this thing, but she went with the boy to the lower valleys.

She was weak and could not travel fast. The cold made her cough. But they went all the way to the place where the deer were in the thickets and he took a deer. They ate deer liver and those parts that make strength for a weak body. They made packs of the meat and they went back to the lodge. It was a long journey and she was tired when they returned. She was so weak she could scarcely stand on her feet. He made her lie on her bed and he cooked soup from the tough winter meat for her. She drank some of the soup, but she was too weak to eat the meat. She had a pain in her chest and she coughed and would not let him see what came out of her throat.

She said, "Help me to sit up," and he helped her. She tried to work at a basket, but her hands would not do what she told them to do. He helped her. He sat beside her and he made the basket for her. She said his hands were better than her hands at this. He said, "My hands have watched your hands many times. My hands are your hands now."

The pain became worse. She was so hot she could not have even one robe around her shoulders. Then she was cold, so cold that he put his red blanket around her and still she shivered. She said, "I am sick." He said, "I am singing the songs for making you well." She said, "I do not hear these songs." He said, "I am singing them inside." She said, "Sing them outside. I am very sick."

He sang the songs all that night. He made soup, but she could not swallow it. She said, "I am not going to get over this sickness, my son." He said, "I will not sing the song for going away yet." She said, "No. Not yet."

All that day he sang the songs for making her well. Late that afternoon she could not hear those songs. She talked about things he did not know. About Pagosa, and the sawmill, and Blue Elk, and Frank No Deer. He heard these things for the first time. Then she did not talk. It was dark; it was night. Then she said, "Sing the song for going away, my son." She tried to reach up and touch his face. Her arm was too weak. He took her hand and held it to his face. Then she died.

He made mourning all that night. When it was daylight, he put her new deerskin dress on her and wrapped her in the red blanket, then in a deerskin robe, so she would be warm on the long journey. He put dried berries and dried meat, the last there was in the lodge, in two baskets. Then he made a drag of a deerskin and he took her up the mountain to the cave in the rocks where they had buried his father. His father was no longer there; only a few bones and a part of a robe and a broken basket. The boy buried his mother there, in the old way, and he sang the old songs. Then he went back down the mountain to the lodge, and he was alone.

LOST SISTER

DOROTHY M. JOHNSON

Our household was full of women, who overwhelmed my uncle Charlie and sometimes confused me with their bustle and chatter. We were the only men on the place. I was nine years old when still another woman came — Aunt Bessie, who had been living with the Indians.

When my mother told me about her, I couldn't believe it. The savages had killed my father, a cavalry lieutenant, two years before. I hated Indians and looked forward to wiping them out when I got older. (But when I was grown, they were no menace any more.)

"What did she live with the hostiles[1] for?" I demanded.

"They captured her when she was a little girl," Ma said. "She was three years younger than you are. Now she's coming home."

High time she came home, I thought. I said so, promising, "If they was ever to get me, I wouldn't stay with 'em long."

Ma put her arms around me. "Don't talk like that. They won't get you. They'll never get you."

I was my mother's only real tie with her husband's family. She was not happy with those masterful women, my aunts Margaret, Hannah, and Sabina, but she would not go back East where she came from. Uncle Charlie managed the store the aunts owned, but he wasn't really a member of the family — he was just Aunt Margaret's husband. The only man who had belonged was my father, the aunts' younger brother. And I belonged, and someday the store would be mine. My mother stayed to protect my heritage.

None of the three sisters, my aunts, had ever seen Aunt Bessie. She had been taken by the Indians before they were born. Aunt Mary had known her — Aunt Mary was two years older — but she lived a thousand miles away now and was not well.

There was no picture of the little girl who had become a legend. When the family had first settled here, there was enough struggle to feed and clothe the children without having pictures made of them.

Even after army officers had come to our house several times and there had been many letters about Aunt Bessie's delivery from the savages, it was a long time before she came. Major Harris, who made the final arrangements, warned my aunts that they would have problems, that Aunt Bessie might not be able to settle down easily into family life.

[1]**hostiles** (hos′ tlz): enemies or unfriendly people.

This was only a challenge to Aunt Margaret, who welcomed challenges. "She's our own flesh and blood," Aunt Margaret trumpeted. "Of course she must come to us. My poor, dear sister Bessie, torn from her home forty years ago!"

The major was earnest but not tactful. "She's been with the savages all those years," he insisted. "And she was only a little girl when she was taken. I haven't seen her myself, but it's reasonable to assume that she'll be like an Indian woman."

My stately aunt Margaret arose to show that the audience[2] was ended. "Major Harris," she intoned, "I cannot permit anyone to criticize my own dear sister. She will live in my home, and if I do not receive official word that she is coming within a month, I shall take steps."

Aunt Bessie came before the month was up.

The aunts in residence made valiant preparations. They bustled and swept and mopped and polished. They moved me from my own room to my mother's — as she had been begging them to do because I was troubled with nightmares. They prepared my old room for Aunt Bessie with many small comforts — fresh doilies everywhere, hairpins, a matching pitcher and bowl, the best towels, and two new nightgowns in case hers might be old. (The fact was that she didn't have any.)

"Perhaps we should have some dresses made," Hannah suggested. "We don't know what she'll have with her."

"We don't know what size she'll take, either," Margaret pointed out. "There'll be time enough for her to go to the store after she settles down and rests for a day or two. Then she can shop to her heart's content."

Ladies of the town came to call almost every afternoon while the preparations were going on. Margaret promised them that as soon as Bessie had recovered sufficiently from her ordeal, they should all meet her at tea.

Margaret warned her anxious sisters, "Now, girls, we mustn't ask her too many questions at first. She must rest for a while. She's been though a terrible experience." Margaret's voice dropped way down with those last two words, as if only she could be expected to understand.

[2]**audience:** *here,* formal hearing or conference with an important person.

Indeed Bessie had been through a terrible experience, but it wasn't what the sisters thought. The experience from which she was suffering, when she arrived, was that she had been wrenched from her people, the Indians, and turned over to strangers. She had not been freed. She had been made a captive.

Aunt Bessie came with Major Harris and an interpreter, a half-blood with greasy black hair hanging down to his shoulders. His costume was half army and half primitive. Aunt Margaret swung the door open wide when she saw them coming. She ran out with her sisters following, while my mother and I watched from a window. Margaret's arms were outstretched, but when she saw the woman closer, her arms dropped and her glad cry died.

She did not cringe, my aunt Bessie who had been an Indian for forty years, but she stopped walking and stood staring, helpless among her captors.

The sisters had described her often as a little girl. Not that they had ever seen her, but she was a legend, the captive child. Beautiful blonde curls, they said she had, and big blue eyes — she was a fairy child, a pale-haired little angel who ran on dancing feet.

The Bessie who came back was an aging woman who plodded in moccasins, whose dark dress did not belong on her bulging body. Her brown hair hung just below her ears. It was growing out; when she was first taken from the Indians, her hair had been cut short to clean out the vermin.

Aunt Margaret recovered herself, and instead of embracing this silent, stolid woman, satisfied herself by patting an arm and crying, "Poor dear Bessie, I am your sister Margaret. And here are our sisters Hannah and Sabina. We do hope you're not all tired out from your journey!"

Aunt Margaret was all graciousness, because she had been assured beyond doubt that this was truly a member of the family. She must have believed — Aunt Margaret could believe anything — that all Bessie needed was to have a nice nap and wash her face. Then she would be as talkative as any of them.

The other aunts were quick-moving and sharp of tongue. But this one moved as if her sorrows were a burden on her bowed shoulders, and when she spoke briefly in answer to the interpreter, you could not understand a word of it.

Aunt Margaret ignored these peculiarities. She took the party

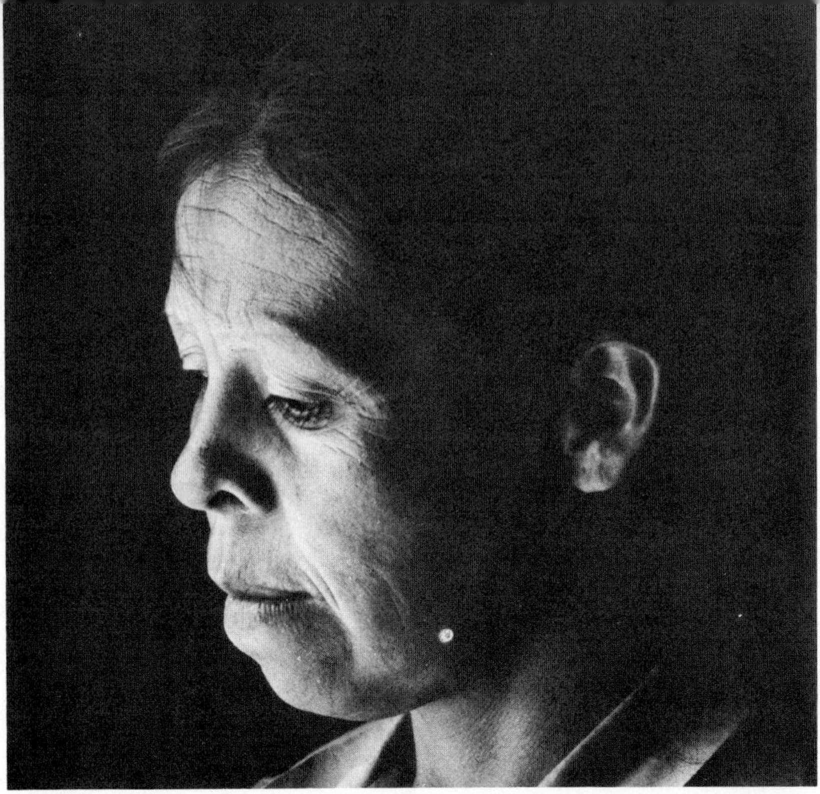

into the front parlor — even the interpreter, when she understood there was no avoiding it. She might have gone on battling with the major about him, but she was in a hurry to talk to her lost sister.

"You won't be able to converse with her unless the interpreter is present," Major Harris said. "Not," he explained hastily, "because of any regulation, but because she has forgotten English."

Aunt Margaret gave the half-blood interpreter a look of frowning doubt and let him enter. She coaxed Bessie. "Come, dear, sit down."

The interpreter mumbled, and my Indian aunt sat cautiously on a needlepoint chair. For most of her life she had been living with the people who sat comfortably on the ground.

The visit in the parlor was brief. Bessie had had her instructions before she came. But Major Harris had a few warnings for the family. "Technically, your sister is still a prisoner," he explained, ignoring Margaret's start of horror. "She will be in your custody. She may walk in your fenced yard, but she must not leave it without official permission.

"Mrs. Raleigh, this may be a heavy burden for you all. But she has been told all this and has expressed willingness to conform to these restrictions. I don't think you will have any trouble keeping her here." Major Harris hesitated, remembered that he was a soldier and a brave man, and added, "If I did, I wouldn't have brought her."

There was the making of a sharp little battle, but Aunt Margaret chose to overlook the challenge. She could not overlook the fact that Bessie was not what she had expected.

Bessie certainly knew that this was her lost white family, but she didn't seem to care. She was infinitely sad, infinitely removed. She asked one question: "Ma-ry?" and Aunt Margaret almost wept with joy.

"Sister Mary lives a long way from here," she explained, "and she isn't well, but she will come as soon as she's able. Dear sister Mary!"

The interpreter translated this, and Bessie had no more to say. That was the only understandable word she ever did say in our house, the remembered name of her older sister.

When the aunts, all chattering, took Bessie to her room, one of them asked, "But where are her things?"

Bessie had no things, no baggage. She had nothing at all but the clothes she stood in. While the sisters scurried to bring a comb and other oddments, she stood like a stooped monument, silent and watchful. This was her prison. Very well, she would endure it.

"Maybe tomorrow we can take her to the store and see what she would like," Aunt Hannah suggested.

"There's no hurry," Aunt Margaret declared thoughtfully. She was getting the idea that this sister was going to be a problem. But I don't think Aunt Margaret ever really stopped hoping that one day Bessie would cease to be different, that she would end her stubborn silence and begin to relate the events of her life among the savages, in the parlor over a cup of tea.

My Indian aunt accustomed herself, finally, to sitting on the chair in her room. She seldom came out, which was a relief to her sisters. She preferred to stand, hour after hour, looking out the window — which was open only about a foot, in spite of all Uncle Charlie's efforts to budge it higher. And she always wore moccasins. She never was able to wear shoes from the store but seemed to treasure the shoes brought to her.

The aunts did not, of course, take her shopping after all. They made her a couple of dresses; and when they told her, with signs and voluble explanations, to change her dress, she did.

After I found that she was usually at the window, looking across the flat land to the blue mountains, I played in the yard so I could stare at her. She never smiled, as an aunt should, but she looked at me sometimes, thoughtfully, as if measuring my worth. By performing athletic feats, such as walking on my hands, I could get her attention. For some reason, I valued it.

She didn't often change expression, but twice I saw her scowl with disapproval. Once was when one of the aunts slapped me in a

casual way. I had earned the slap, but the Indians did not punish children with blows. Aunt Bessie was shocked, I think, to see that white people did. The other time was when I talked back to someone with spoiled, small-boy insolence — and that time the scowl was for me.

The sisters and my mother took turns, as was their Christian duty, in visiting her for half an hour each day. Bessie didn't eat at the table with us — not after the first meal.

The first time my mother took her turn, it was under protest. "I'm afraid I'd start crying in front of her," she argued, but Aunt Margaret insisted.

I was lurking in the hall when Ma went in. Bessie said something, then said it again, peremptorily, until my mother guessed what she wanted. She called me and put her arm around me as I stood beside her chair. Aunt Bessie nodded, and that was all there was to it.

Afterward, my mother said, "She likes you. And so do I." She kissed me.

"I don't like her," I complained. "She's queer."

"She's a sad old lady," my mother explained. "She had a little boy once, you know."

"What happened to him?"

"He grew up and became a warrior. I suppose she was proud of him. Now the army has him in prison somewhere. He's half Indian. He was a dangerous man."

He was indeed a dangerous man, and a proud man, a chief, a bird of prey whose wings the army had clipped after bitter years of trying.

However, my mother and my Indian aunt had that one thing in common: they both had sons. The other aunts were childless.

There was a great to-do about having Aunt Bessie's photograph taken. The aunts, who were stubbornly and valiantly trying to make her one of the family, wanted a picture of her for the family album. The government wanted one too, for some reason — perhaps because someone realized that a thing of historic importance had been accomplished by recovering the captive child.

Major Harris sent a young lieutenant with the greasy-haired interpreter to discuss the matter in the parlor. (Margaret, with great foresight, put a clean towel on a chair and saw to it the interpreter sat there.) Bessie spoke very little during that meeting, and of course we understood only what the half-blood *said* she was saying.

No, she did not want her picture made. No.

But your son had his picture made. Do you want to see it? They teased her with that offer, and she nodded.

If we let you see his picture, then will you have yours made?

She nodded doubtfully. Then she demanded more than had been offered: If you let me keep his picture, then you can make mine.

No, you can only look at it. We have to keep his picture. It belongs to us.

My Indian aunt gambled for high stakes. She shrugged and spoke, and the interpreter said, "She not want to look. She will keep or nothing."

My mother shivered, understanding as the aunts could not understand what Bessie was gambling — all or nothing.

Bessie won. Perhaps they had intended that she should. She was allowed to keep the photograph that had been made of her son. It has been in history books many times — the half-white chief, the valiant leader who was not quite great enough to keep his Indian people free.

His photograph was taken after he was captured, but you would never guess it. His head is high, his eyes stare with boldness but not with scorn, his long hair is arranged with care — dark hair braided on one side and with a tendency to curl where the other side hangs loose — and his hands hold the pipe like a royal scepter.

That photograph of the captive but unconquered warrior had its effect on me. Remembering him, I began to control my temper and my tongue, to cultivate reserve as I grew older, to stare with boldness but not scorn at people who annoyed or offended me. I never met him, but I took silent pride in him — Eagle Head, my Indian cousin.

Bessie kept his picture on her dresser when she was not holding it in her hands. And she went like a docile, silent child to the photograph studio, in a carriage with Aunt Margaret early one morning, when there would be few people on the street to stare.

Bessie's photograph is not proud, but pitiful. She looks out with no expression. There is no emotion there, no challenge, only the face of an aging woman with short hair, only endurance and patience. The aunts put a copy in the family album.

But they were nearing the end of their tether.[3] The Indian aunt was a solid ghost in the house. She did nothing because there was nothing for her to do. Her gnarled hands must have been skilled at squaws' work, at butchering meat and scraping and tanning hides, at making tepees and beading ceremonial clothes. But her skills were useless and unwanted in a civilized home. She did not even sew when my mother gave her cloth and needles and thread. She kept the sewing things beside her son's picture.

She ate (in her room) and slept (on the floor) and stood looking out the window. That was all, and it could not go on. But it had to go on, at least until my sick Aunt Mary was well enough to travel — Aunt Mary, who was her older sister, the only one who had known her when they were children.

The sisters' duty visits to Aunt Bessie became less and less visits and more and more duty. They settled into a bearable routine. Margaret had taken upon herself the responsibility of trying to make Bessie talk. Make, I said, not teach. She firmly believed that her stubborn and unfortunate sister needed only encouragement from a strong-willed person. So Margaret talked, as to a child, when she bustled in:

"Now there you stand, just looking, dear. What in the world is there to see out there? The birds — are you watching the birds? Why don't you try sewing? Or you could go for a little walk in the yard. Don't you want to go out for a nice little walk?"

Bessie listened and blinked.

Margaret could have understood an Indian woman's not being able to converse in a civilized tongue, but her own sister was not an Indian. Bessie was white; therefore she should talk the language her sisters did — the language she had not heard since early childhood.

Hannah, the put-upon aunt, talked to Bessie too, but she was delighted not to get any answers and not to be interrupted. She bent over her embroidery when it was her turn to sit with Bessie and told her troubles in an unending flow. Bessie stood looking out the window the whole time.

Sabina, who had just as many troubles, most of them emanating[4] from Margaret and Hannah, went in like a martyr, firmly

[3]**tether:** *here,* endurance. A tether is a rope or chain fastening an animal.
[4]**emanating** (em'ə·nāt·ing): coming forth.

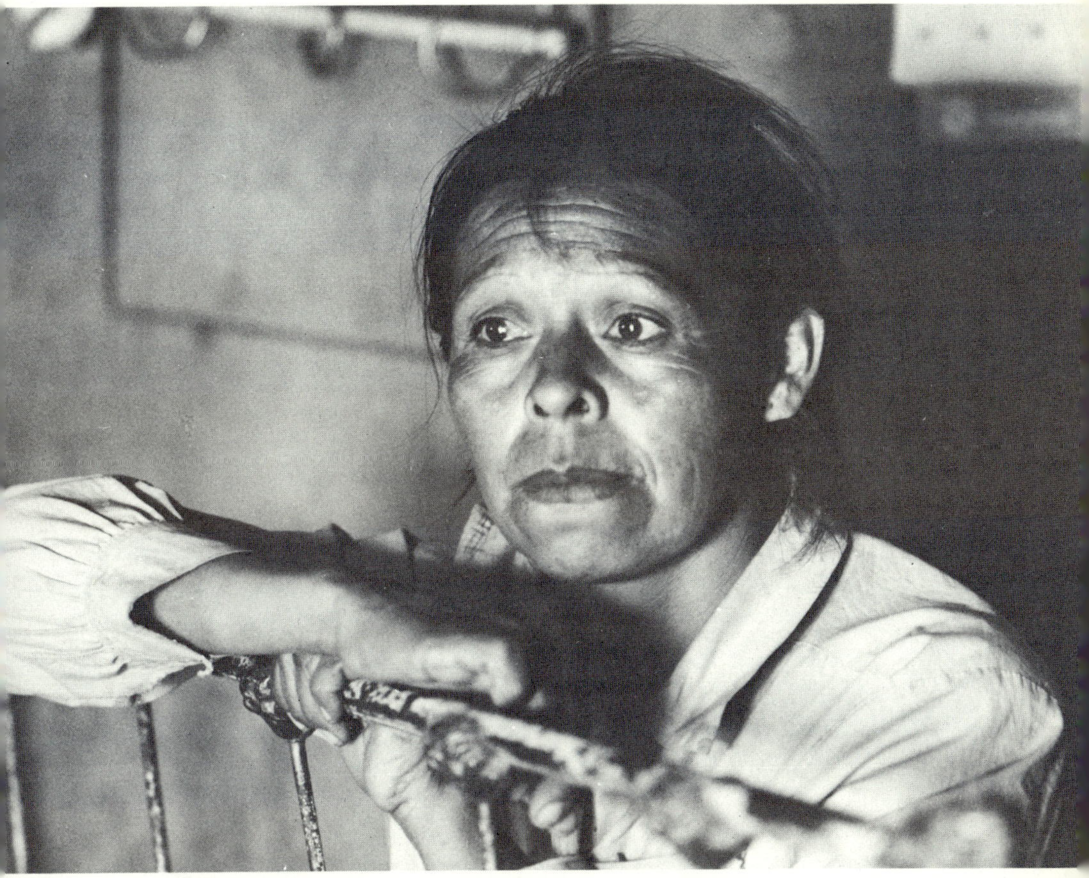

clutching her Bible, and read aloud from it until her time was up. She took a small clock along so that she would not, because of annoyance, be tempted to cheat.

After several weeks Aunt Mary came, white and trembling and exhausted from her illness and the long, hard journey. The sisters tried to get the interpreter in but were not successful. (Aunt Margaret took that failure pretty hard.) They briefed Aunt Mary, after she had rested, so the shock of seeing Bessie would not be too terrible. I saw them meet, those two.

Margaret went to the Indian woman's door and explained volubly who had come, a useless but brave attempt. Then she stood aside, and Aunt Mary was there, her lined white face aglow, her arms outstretched. "Bessie! Sister Bessie!" she cried.

And after one brief moment's hesitation, Bessie went into her arms, and Mary kissed her sun-dark, weathered cheek. Bessie spoke. "Ma-ry," she said. "Ma-ry." She stood with tears running down her face and her mouth working. So much to tell, so much suffering and fear — and joy and triumph, too — and the sister there at last who might legitimately hear it all and understand.

But the only English word that Bessie remembered was *Mary,* and she had not cared to learn any others. She turned to the dresser, took her son's picture in her work-hardened hands, reverently, and held it so her sister could see. Her eyes pleaded.

Mary looked on the calm, noble, savage face of her half-blood nephew and said the right thing: "My, isn't he handsome!" She put her head on one side and then the other. "A fine boy, sister," she approved. "You must" — she stopped, but she finished — "be awfully proud of him, dear!"

Bessie understood the tone if not the words. The tone was admiration. Her son was accepted by the sister who mattered. Bessie looked at the picture and nodded, murmuring. Then she put it back on the dresser.

Aunt Mary did not try to make Bessie talk. She sat with her every day for hours, and Bessie did talk — but not in English. They sat holding hands for mutual comfort while the captive child, grown old and a grandmother, told what had happened in forty years. Aunt Mary said that was what Bessie was talking about. But she didn't understand a word of it and didn't need to.

"There is time enough for her to learn English again," Aunt Mary said. "I think she understands more than she lets on. I asked her if she'd like to come and live with me, and she nodded. We'll have the rest of our lives for her to learn English. But what she has been telling me — she can't wait to tell that. About her life, and her son."

"Are you sure, Mary dear, that you should take the responsibility of having her?" Margaret said dutifully, no doubt shaking in her shoes for fear Mary would change her mind now that deliverance was in sight. "I do believe she'd be happier with you, though we've done all we could."

Margaret and the older sisters would certainly be happier with Bessie somewhere else. And so, it developed, would the United States government.

Major Harris came with the interpreter to discuss details, and they told Bessie she could go, if she wished, to live with Mary a thousand miles away. Bessie was patient and willing, stolidly agreeable. She talked a great deal more to the interpreter than she had ever done before. He answered at length and then explained to the others that she had wanted to know how she and Mary would travel to this far country. It was hard, he said, for her to understand just how far they were going.

Later we knew that the interpreter and Bessie had talked about much more than that.

Next morning, when Sabina took breakfast to Bessie's room, we heard a cry of dismay. Sabina stood holding the tray, repeating, "She's gone out the window! She's gone out the window!"

And so she had. The window that had always stuck so that it would not raise more than a foot was open wider now. And the photograph of Bessie's son was gone from the dresser. Nothing else was missing except Bessie and the decent dark dress she had worn the day before.

My uncle Charlie got no breakfast that morning. With Margaret shrieking orders, he leaped on a horse and rode to the telegraph station.

Before Major Harris got there with half a dozen cavalrymen, civilian scouts were out searching for the missing woman. They were expert trackers. Their lives had depended, at various times, on their ability to read the meaning of a turned stone, a broken twig, a bruised leaf. They found that Bessie had gone south. They tracked her for ten miles. And then they lost the trail, for Bessie was as skilled as they were. Her life had sometimes depended on leaving no stone or twig or leaf marked by her passage. She traveled fast at first. Then, with time to be careful, she evaded the followers she knew would come.

The aunts were stricken with grief — at least Aunt Mary was — and bowed with humiliation about what Bessie had done. The blinds were drawn, and voices were low in the house. We had been pitied because of Bessie's tragic folly in having let the Indians make a savage of her. But now we were traitors because we had let her get away.

Aunt Mary kept saying pitifully, "Oh, why did she go? I thought she would be contented with me!"

The others said that it was, perhaps, all for the best. Aunt Margaret proclaimed, "She has gone back to her own." That was what they honestly believed, and so did Major Harris.

My mother told me why she had gone. "You know that picture she had of the Indian chief, her son? He's escaped from the jail he was in. The fort got word of it, and they think Bessie may be going to where he's hiding. That's why they're trying so hard to find her. They think," my mother explained, "that she knew of his escape before they did. They think the interpreter told her when he was here. There was no other way she could have found out."

They scoured the mountains to the south for Eagle Head and Bessie. They never found her, and they did not get him until a year later, far to the north. They could not capture him that time. He died fighting.

After I grew up, I operated the family store, disliking storekeeping a little more every day. When I was free to sell it, I did, and went to raising cattle. And one day, riding in a canyon after strayed steers, I found — I think — Aunt Bessie. A cowboy who worked for me was along, or I would never have let anybody know.

We found weathered bones near a little spring. They had a mystery on them, those nameless human bones suddenly come upon. I could feel old death brushing my back.

"Some prospector," suggested my riding partner.

I thought so too until I found, protected by a log, sodden scraps of fabric that might have been a dark, respectable dress. And wrapped in them was a sodden something that might have once been a picture.

The man with me was young, but he had heard the story of the captive child. He had been telling me about it, in fact. In the passing years it had acquired some details that surprised me. Aunt Bessie had become once more a fair-haired beauty, in this legend that he had heard, but utterly sad and silent. Well, sad and silent she really was.

I tried to push the sodden scrap of fabric back under the log, but he was too quick for me. "That ain't no shirt, that's a dress!" he announced. "This here was no prospector — it was a woman!" He paused and then announced with awe, "I bet you it was your Indian aunt!"

I scowled and said, "Nonsense. It could be anybody."

He got all worked up about it. "If it was *my* aunt," he declared, "I'd bury her in the family plot."

"No," I said, and shook my head.

We left the bones there in the canyon, where they had been for forty-odd years if they were Aunt Bessie's. And I think they were. But I would not make her a captive again. She's in the family album. She doesn't need to be in the family plot.

If my guess about why she left us is wrong, nobody can prove it. She never intended to join her son in hiding. She went in the opposite direction to lure pursuit away.

What happened to her in the canyon doesn't concern me, or anyone. My aunt Bessie accomplished what she set out to do. It was not her life that mattered, but his. She bought him another year.

Indian Sky

ALFRED KREYMBORG

The old squaw
is one
with the old stone behind her.
Both have squatted there —
ask mesa[1]
or mountain how long?
. . . the bowl she holds —
clay ritual of her faith —
is one
with the thought of the past;
and one with the now;
. . . the earth holds her
as she holds the bowl —
ask kiva[2]
or shrine how much longer?
No titan[3]
or destroyer,
or future thought,
can part
earth and this woman,
woman and bowl:
the same shawl
wraps them around.

[1]**mesa** (mā′sə): small, high plateau.

[2]**kiva** (kē′və): underground room used for Pueblo ceremonies or councils.

[3]**titan** (tīt′n): person or thing having enormous size, strength, or power.

Bullet Removed
Kennedy
LBJ Orders
Guard for
Candidates
Triumph Was Brief

To what extent should a person's survival depend upon his beliefs?

Not to Go with

In the third year of the war, Frantizek Zaremski was arrested by the invaders on a charge of spreading underground literature — specifically, for carrying about his person a poem a friend from Gdynia[1] had given him, which began:

Sleep, beloved Hitler, planes will come by night. . . .

After he had spent six weeks of a three-year sentence for this crime in the Gestapo[2] prison at Inowroczon, Zaremski was sent to Kalice to do carpentry. By bad luck, at the time when his term expired, the Russians had broken through at the Vistula, and his captors, instead of releasing him, took him, in their general panic, to the transfer camp for Polish political prisoners at Rodogoszez, where he was placed in Hall Number Four with nine hundred men. Altogether there were between two and three thousand men and women — no Jews, only "Aryan"[3] Poles suspected or convicted of political activity — in the prison.

Late in the evening of Wednesday, January 17, 1945, three days before Lodz was to fall to the Russians, all the prisoners were gathered on the third and fourth floors of the main building, even those who were sick, and there they all lay down on wooden bunks and floors to try to sleep. At about two in the morning guards came and ordered the inmates to get up for roll call.

They divided the prisoners into groups of about twenty each and lined up the groups in pairs. Zaremski was in the second group. SS[4] men led it down concrete stairs in a brick-walled stairwell at

[1]**Gdynia:** city in Poland, where the story takes place, during World War II.

[2]**Gestapo** (gə·stä′pō): German internal security police under the Nazi regime.

[3]**"Aryan"** (âr′ē·ən): *here,* non-Jewish persons of Indo-European descent.

[4]**SS:** abbreviation for Schutzstaffel, the elite military and police unit of the Nazi Party.

the Others

JOHN HERSEY

one end of the building and halted it on a landing of the stairway, near a door opening into a large loft on the second floor. The first group had apparently been led down to the ground floor.

Someone gave an order that the prisoners should run in pairs into the loft as fast as they could. When the first pairs of Zaremski's group ran in, SS men with their backs to the wall inside the room began to shoot at them from behind. Zaremski's turn came. He ran in terror. A bullet burned through his trouser leg. Another grazed his thigh. He fell down and feigned death.

Others, from Zaremski's and later groups, ran into the hall and were shot and fell dead or wounded on top of Zaremski and those who had gone first. At one time Zaremski heard the Polish national anthem being sung somewhere.

Finally the running and shooting ended, and there ensued some shooting on the upper floors, perhaps of people who had refused to run downstairs.

SS men with flashlights waded among the bodies, shining lights in the faces of the prostrate victims. Any wounded who moaned or moved, or any whose eyes reacted when the shafts of light hit their faces, were dispatched with pistol shots. Somehow Zaremski passed the test of pretense.

As dawn began to break, Zaremski heard the iron doors of the main building being locked, and he heard some sort of grenades or bombs being thrown into the lowest hall and exploding there; they seemed to him to make only smoke, but they may have been incendiaries.[5] Later, in any case, the ground floor began to burn. Perhaps benzine or petrol had been poured around. Zaremski was still lying among the bodies of others.

[5]**incendiaries** (in·sen'dē·er·ēz): fire bombs.

There were several who were still alive, and they began jumping out of the burning building, some from windows on the upper stories. A few broke through a skylight to the roof, tied blankets from the prisoners' bunks into long ropes, and let themselves down outside. Zaremski, now scurrying about the building, held back to see what would happen. Those who jumped or climbed down were shot at leisure in the camp enclosure by SS men in the turrets on the walls, and Zaremski decided to try to stay inside.

On the fourth floor, at the top of the reinforced concrete staircase, in the bricked stairwell at the end of the building, Zaremski found the plant's water tank, and for a time he and others poured water over the wounded lying on the wooden floors in the main rooms. Later Zaremski took all his clothes off, soaked them in the tank, and put them back on. He lay down and kept pouring water over himself. He put a soaked blanket around his head.

The tank was a tall one, separated from the main room by the stairwell's brick wall, and when the fire began to eat through the wooden floor of the fourth story and the heat in the stairwell grew unbearable, Zaremski climbed up and got right into the water in the

tank. He stayed immersed there all day long. Every few minutes he could hear shots from the wall turrets. He heard floors of the main halls fall and heard the side walls collapse. The staircase shell and the concrete stairs remained standing.

It was evening before the shooting and the fire died down. When he felt sure both had ended, Zaremski pulled himself out of the tank and lay awhile on the cement floor beside it. Then, his strength somewhat restored, he made his way down the stairs, and on the way he found six others who were wounded but could walk.

The seven went outside. Dusk. All quiet. They thought the Germans had left, and they wanted to climb the wall and escape. The first three climbed up and dropped away in apparent safety, but then the lights flashed on in the turrets and bursts of firing broke out. Three of the remaining four decided to take their chances at climbing out after total darkness; they did not know whether the first three had been killed or had escaped. Only Zaremski decided to stay.

The three climbed, but this time the lights came sooner, and the guards killed all three while they were still scaling the wall.

Zaremski crept into the camp's storehouse in a separate building. Finding some damp blankets, he wrapped them around himself and climbed into a big box, where he stayed all night. Once during the night he heard steps outside the building, and in the early morning he heard walking again. This time the footsteps approached the storeroom door. The door opened. The steps entered. Through the cracks of the box Zaremski sensed that the beam of a flashlight was probing the room. Zaremski could hear box tops opening and slamming and a foot kicking barrels. He held the lid of his box from the inside. Steps came near, a hand tried the lid, but Zaremski held tight, and the searcher must have decided the box was locked or nailed down. The footsteps went away.

Later two others came at different times and inspected the room, but neither tried Zaremski's box; the third hunter locked the door from the outside.

Much later Zaremski heard a car start and drive away.

Much later still — some time on the nineteenth of January in the year of victory — Zaremski heard the Polish language being spoken, even by the voices of women and children. He jumped out of the box and broke the window of the storehouse and climbed out to his countrymen.

Ashes for the Wind

HERNANDO TELLEZ

The man had a cordially sinister air. He had been trying unsuccessfully for at least half an hour to say what he had to say. He was seated on a thick log by the door of the house. He had not taken off his greasy hat, a cheap brown felt, and he kept his eyes on the ground while he talked. Juan knew him well. He was the son of Simón Arévalo and his wife, Laura. Even as a boy he had had a bad name, but nobody would have dreamed that he would do what they now said he was doing to old friends of his parents in the vicinity. Juan had not believed it, but now. . . . "It would be best for you to clear out," the man repeated without lifting his eyes from the ground. Juan made no answer. The day had dawned lowering, and the leaden clouds threatened heavy rain. The air was sultry. Juan looked beyond the visitor's hat, out over the fields, green, yellow, yellow-ripe, green again, a brighter green, and then paler. There was a good view of the valley from where he was standing. It was a good spot from which to see the green heads of the grain rippling in the wind.

"Who's there?" came the clear high-pitched voice of his wife from the kitchen. He did not answer. The visitor sat on, his head lowered. With one of his dusty shoes he kept scuffing a little mound of fine dust against the other shoe and then carefully pressing it flat with the sole. "You'd better clear out," he said again, this time lifting his face. Juan looked at him. And thought to himself how much he looked like his father except for his eyes, which were the color of tobacco-leaf, like his mother's.

"Who's there?" This time his wife's voice sounded nearer. And then Carmen was standing in the door that opened on the front porch, holding the baby in her arms. The man got up from the log and mechanically rubbed one of his hands on the seat of his pants. Then he took off his hat. Thick black matted hair sprang up as though released from confinement. It looked as though it had not seen a comb for a long time. "Good morning, Señora Carmen," said the visitor. The baby was playing with his mother's throat, trying to bury his fingers in its softness. He was only a few months old, and he reeked of mother's milk and soiled diapers.

Juan said nothing. The man was visibly disconcerted. For a few seconds there was no sound but the silence of the countryside, and in the midst of this silence, the ever-merged, ever-latent[1] noises of nature. The valley throbbed intact under the sullen morning. "But the sun is going to come out," Juan thought.

"Well, I'll be going," the visitor said, and the men said good-bye. Carmen was silent, her eyes on her husband. The man put on his hat again, turned his back on them, and walked slowly to the fence gate, some ten or fifteen yards from the house, and opened it carefully, the unoiled hinges giving their customary creak despite his care. Ordinary hinges made by the local blacksmith.

"They'd better clear out." Why? The son of Simón Arévalo and his dead wife, Laura, had tried to explain the reasons for nearly half an hour. But he had been so rattled. All this business of the authorities and politics was always complicated. And Simón Arévalo's son was not too clear about it in his own mind, in spite of the fact that he was now hand in glove with the authorities, doing their

[1]**latent** (lāt'nt): hidden.

dirty work for them. "The scoundrel," thought Juan. "He said if we weren't gone by the end of the week, they'd come and put us out." "They'll have to kill us," Carmen answered. "That's what I told him," Juan replied, his face a mask of gloom. They said no more. Carmen went to the kitchen with the child clasped in her arms, and Juan remained alone, standing like a tree before his house.

The neighborhood was a poor one, and there seemed to be no good reason why the authorities should concern themselves with Juan's house and the fields around it. It was of no use to them — a few patches of corn, some hills of potatoes, a vegetable garden with a little stream of water flowing through it, thanks be to God, as Carmen said, from the really fine big place of the Hurtados. As for the house, it was half cabin, half house. Juan was thinking that if the authorities took it from him, they would have to pay the balance of the money he had borrowed some years back to build the kitchen and put in the septic tank. But could it be true that they would have to clear out, as Simón Arévalo's son had told him? Sure, he had voted in the last election. So what? Who hadn't, some for one side, some for the other? And no hard feelings. There always had to be a winner and a loser. Juan let out a laugh. "He was trying to scare me." But no, that wasn't it. He recalled that when he had been in town a week before, there had been something queer. Some of the police, in addition to a gun, had carried a whip. The gun, all right, but why the whip? It puzzled him. The law with a whip in its hand frightened him. Besides, he had noticed something strange in the people. At Don Romulo Linares's store they had refused to sell him oil. They told him they were out of it. But the oil was there, thick and shiny, dripping from the black drum into the funnel, and from the funnel into a bottle behind the counter. He did not say anything, because of the way Don Romulo scowled at him, and he didn't like to have words with anybody. Four policemen were walking around the market place, but there were not many people. He bought a few things — a clay cooking-pot, a bar of soap, a pair of sandals. Then he went into the drugstore for a jar of perfumed vaseline and a roll of cotton. Benavides, the proprietor, asked him, pleasantly but with a mysterious air: "Nothing happened out your way yet?" And as Juan was opening his mouth to answer, Benavides made him a sign to keep quiet. A policeman was walking in, and following him came Simón Arévalo's son. The policeman rapped on the wooden counter with his whip. Benavides lost color and quickly wrapped up Juan's purchases. "What's going on here?" asked the policeman. Arévalo recognized Juan, but he looked at him as though he had never seen Juan before. The policeman did not give Benavides time to answer. Turning to Juan, and switching the whip against his own pants, he said to him: "So you, too, are one of those who are resisting?" Juan must have gone pale like Benavides, for he felt his heart begin to thud. He would have liked to slap the policeman's face; just because he was a policeman he had no right to talk like that to a peaceable man buying a jar of vaseline and a roll of cotton in the drugstore and bothering nobody. Arévalo spoke up: "Yes, he's one of the Reds; he lives near here, over Tres Espigas way." Juan stood as though rooted to the floor, his eyes riveted on the short wooden whipstock with a hole at one end through which the leather thongs were strung. The knotty wood looked like a long finger whose joints were swollen with rheumatism. And the whip went on swishing against the coarse, khaki-colored cloth of the uniform. "M-hm, m-hm," growled the policeman meaningly. "But he's one of the quiet ones; I know him," put in Arévalo. The beat of the whip against the cloth stopped. "We'll see, we'll see . . . " and a sardonic[2] smile creased the policeman's face. "There'll be no more haw-hawing around here, you hear, Benavides. And that goes for you, too. . . ." They left.

Juan's mouth felt dry. He picked up the package from the counter, fished in his pocket for the forty-five centavos he owed, and said good-bye to Benavides, whose hands were still shaking and who was as pale as a person with a sudden attack of cramps.

Now the threat had taken form in the person of the son of Simón Arévalo. Juan recalled that Simón Arévalo had been his friend, and that this boy had not seemed really bad. It was just that he liked to go around shooting off his mouth about this business of politics. But what was he up to now? If he had joined the force, that would be clear. But he wasn't wearing a uniform. Ever since things had become hot, Arévalo had been hand in glove with the authorities. People in town said that he was always in the mayor's office or drinking with the police. An informer — that's what he was, and an informer who had the advantage of knowing everybody for five, maybe ten leagues around. Why wouldn't he? He had been born right here, like Simón, his father, and like his grandfather. Why wouldn't he, when he had gone to the village school, barefooted like Juan, and like him had run all over, knowing the names of all the owners and renters and tenants and hired hands, working here and there till he was grown up, had put on shoes and a felt hat, and had settled down in the neighborhood?

The shots awakened Carmen first, then Juan; then the baby began to cry. It was getting light, and the objects in the room were clearly visible. As he jumped out of bed, Juan calculated what time it would be: about five. The shots were repeated, this time closer. He finished pulling on his pants, buckled his belt, and rushed to the door. He had been right about the time; the sky was shedding a milky light over the fields. "Yes, it's five. It's going to be a nice day," he thought without realizing it. The hinges of the fence gate announced that someone was coming in. Two men walked

[2]**sardonic** (sär·don'ik): bitter; scornful; mocking.

through. Juan recognized them at once; one was Arévalo, the other the policeman with the whip he had seen at Benavides's drugstore. Then Arévalo's warning had come true? It was twelve days since his visit. Everything had worked out just as he had said. "A week, be out of here in a week. That's the best thing you can do. Otherwise. . . . " Now here was Arévalo again, but this time with the law.

The policeman fired another shot into the air as he approached Juan. "Sounds good, doesn't it?" he said, "and you'll hear a lot more tomorrow at this same time if you're not away from here. You understand?" He cocked his pistol again, taking aim at the slender cornstalks, just for the fun of it. Arévalo stood there with hanging head. He did not look at Juan or at Carmen, who had come running to see what was happening. "You've had fair warning. Clear out, and clear out fast." The policeman put the pistol in his holster, took Arévalo by the arm, and walked away. It was only then that Juan realized that the policeman's breath reeked of brandy.

Everyone did his duty, Arévalo and the law, Juan and Carmen and the baby. The house burned easily, with a gay crackle of dry thatch, seasoned wood, and old furniture. For two hours, or maybe three. And as a fresh breeze was blowing from the north, whipping the flames, it looked like fair time in the village square. A huge Roman candle. The whip-carrying policeman was having the time of his life, enjoying it much more than his four companions and Arévalo, who had come along to bear witness to whether Juan Martínez had cleared out or put up a resistance.

When they got back to town, they stopped at Linares's store. The mayor was there, leaning lazily against the sacks of corn.

"How did things go?" "All right, Your Honor," Arévalo anwered briefly. "Had Martinez left?" "No," said the policeman. "The damn fool locked the doors and stayed in the house. You understand, we didn't have time to waste. . . . "

The oil went drip-drip from the drum to the funnel and from the funnel to the bottle.

THE TINY CLOSET

WILLIAM INGE

Characters

Mr. Newbold
Mrs. Crosby
Elsie
Mrs. Hergesheimer

Scene: a boarding-rooming house somewhere in a Midwestern city. On stage we see the big living room of the house, which is Victorian in design, with ornate woodwork and high ceiling. The furnishings, too, are Victorian. An ornate wooden stairway is at the Right. The outside entrance is Down Right. When the curtain is up, MR. NEWBOLD *is seen coming down the stairs. He is a man of about fifty, a large man and rather nice-looking. He is always impeccably dressed in the most conservative clothes, a dark blue suit, white shirt, modest tie, a high shine on his black shoes, and his thinning hair carefully combed. He is the sort of man who takes great pride in his grooming. Something now seems to be bothering him. When he gets to the bottom of the stairs, he stops a moment and thinks. Then he calls his landlady.*

Mr. Newbold. Mrs. Crosby! [*No response.*] Mrs. Crosby!

Mrs. Crosby [*coming from the kitchen*]. Yes, Mr. Newbold. I was just straightening up after breakfast. I'm afraid none of the guests like the new bacon I got. It's just as expensive as the other bacon I was serving. On my word it is. Every bit. I didn't buy it to bring down expenses. Not at all. It's *fine* bacon. Hickory smoked. Only it's the kind you slice yourself. That's why I got it. You can cut yourself a good thick slice of bacon that you can really get your teeth into, instead of that other stuff that shrivels up like tissue paper . . . . [*She is the sort of woman who continues talking until someone stops her.*]

Mr. Newbold. Mrs. Crosby . . . .

Mrs. Crosby. Yes, Mr. Newbold?

Mr. Newbold. Mrs. Crosby, you remember, before I moved in, I specified that no one was to enter my closet . . . .

Mrs. Crosby. Indeed I remember, Mr. Newbold. And I told you, you could have your own lock on the closet, like you asked me. No one around here has any keys to your closet but *you,* Mr. Newbold.

Mr. Newbold. Nevertheless, Mrs. Crosby . . . .

Mrs. Crosby. And I told the cleaning woman, "Elsie, you're not to enter Mr. Newbold's closet. Mr. Newbold," I told her, "is a perfectly orderly gentleman who is perfectly capable of keeping his own closet, and you're not to bother it." I gave her strict orders, Mr. Newbold.

Mr. Newbold. Nevertheless, Mrs. Crosby . . . .

Mrs. Crosby. And as for me, goodness, I never go near the rooms. I got enough on my hands downstairs without bothering about the upstairs. I leave all that to Elsie. I don't suppose I been upstairs now in almost . . . .

Mr. Newbold. Mrs. Crosby, a closet is a very small space. That's all I ask in this life. That's all I ask, is just that tiny closet to call my own, my very own.

Mrs. Crosby. I quite understand, Mr. Newbold. We all have to have some place that's private to us, where we don't invite the world to see. I quite understand.

Mr. Newbold. But someone has been monkeying with the lock on the door, Mrs. Crosby.

Mrs. Crosby [*shocked*]. You don't say!

Mr. Newbold. I *do* say, Mrs. Crosby. Someone has been monkeying with that expensive Yale lock I had put on the door.

Mrs. Crosby. Do you suppose someone coulda got up there when no one was around and . . . .

Mr. Newbold. I'm sure I don't know, but I won't stand for anyone's monkeying with that lock, Mrs. Crosby. That room, while I rent it, is my private property, and I gave strict instructions that no one was to go near the closet, and I expect my orders to be respected.

Mrs. Crosby. Of course, Mr. Newbold; you're my favorite of all the roomers. Oh, I wish they was all like you. You keep your room spotless, and you're always so correct around the house. My, you're a model guest. You really are. You should open up a class out here at that night school they have for adults and teach 'em how to behave in their rooming houses. The landladies in this town would get together and thank you.

Mr. Newbold. Thank you, Mrs. Crosby. Nevertheless, I must repeat that that closet is my personal property. There is nothing inside I am ashamed of. It's not that. It's only that I have *some* place, just some little place, that's completely private, that no one has access to. That's all I ask, Mrs. Crosby.

Mrs. Crosby. And I quite understand, Mr. Newbold.

Mr. Newbold. Very well. As long as that is clearly understood, I hope I'll not have to bring the subject up again, Mrs. Crosby.

Mrs. Crosby. I'll tell Elsie again, Mr. Newbold. I'll give special emphasis that no one is to go near that closet.

Mr. Newbold. Thank you, Mrs. Crosby. [*Looks at his watch.*] Goodness! I mustn't dally another minute.

Mrs. Crosby. Did you see Mrs. Hergesheimer in your store yesterday, Mr. Newbold?

Mr. Newbold. Mrs. Hergesheimer? Oh, yes, I believe I did.

Mrs. Crosby. She called me yesterday and told me she seen you. "Whata fine man that Mr. Newbold is!" she says. "You're a lucky woman, Mrs. Crosby, to have such a fine guest. All I got in my house is a lotta old-maid schoolteachers, and I'm always cleanin' up after 'em." That's what she says. "Oh, men are much tidier than women," I told her. I kept schoolteachers once, and they was always in the bathroom, curling their hair in there or giving themselves shampoos, or shaving their legs on my nice bedspreads. *Mr.* Crosby was alive then, and it almost drove him crazy. He never could get to the bathroom when he wanted to, because of them schoolteachers.

Mr. Newbold. Good day, Mrs. Crosby. [*He puts on his hat and goes out the door.*]

Mrs. Crosby. Good day, Mr. Newbold! [*Calling to him outside the door.*] You be here for dinner tonight, won't you?

Mr. Newbold [*off*]. I intend to be.

Mrs. Crosby [*closing the door*]. Good day, Mr. Newbold.

[*Now she comes back inside the room. Curiosity is about to kill her. She is childishly excited. She goes back to the door and peers out to make sure that* MR. NEWBOLD *is on his way. Then she comes back inside the room and goes to the telephone, dialing her number.*]

Mrs. Crosby. Mrs. Hergesheimer? Have you got a minute? He's just left. "You're not to go into my room," he says to me again, for the two-hundredth time. "Someone's been playing with the lock on my door, and you're not to go near that closet," he says, like it was his house and he was ordering me about. Can you beat it? Now what do you suppose he's got hid away in that closet? [*She listens a while.*] No, I can't think he'd be a Communist, Mrs. Hergesheimer. Of course, he *might* be. You never can tell. But I don't think that's it, somehow. [*She listens again.*] Love letters? But why would he need a whole closet for his love letters? — If he's got any. Besides, I don't think he's the type of man that has love letters. [*She listens.*] It's certainly a mystery. I confess, it's certainly a mystery. What would *you* do, Mrs. Hergesheimer? [*A pause.*] You would? Well, hurry over, why don't you, and we'll try again. Hurry! [*She hangs up, then calls out to the kitchen.*] Elsie, you're not to go upstairs for a while. Mrs. Hergesheimer is comin' over, and she and I have some things we wanta do up there, so you're to keep busy with the washin'. Understand?

Elsie [*off*]. Yes, Mrs. Crosby.

[*Now* MRS. CROSBY *spends a few minutes of nervous activity. Her conscience bothers her some, but primarily she is afraid of being caught. She looks out the door again, then out of each window, then calls out to the kitchen again.*]

Mrs. Crosby. Remember, Elsie, you're not to bother me for a while. I'll let you know when I want you, Elsie. You're to stay out there till I call you.

[*Now* MRS. HERGESHEIMER *hurries into the house.*]

Mrs. Hergesheimer. I think you've got every *right,* Mrs. Crosby. Every right.

Mrs. Crosby. That's what I've been telling myself, Mrs. Hergesheimer. I've got every right. For all I know, I may be harboring a *spy,* or a criminal, or a lunatic. What's he got in that closet that he don't want anyone to see? Can you tell me? It must be something he's ashamed of, or he wouldn't mind if anyone saw. Isn't that what you say? And if it's something he's ashamed of, I think we should find out what it is. You can't tell — he might have a bomb in there he meant to destroy us with. I'm not gonna set idly by while someone is plotting something, Mrs. Hergesheimer. I pride myself I'm a real American, and I say if anyone's got any secrets he wants to keep hid, let 'm come out into the open and declare himself. Mr. Newbold has always seemed like a fine man, and I got nothin' against him personally, and he's the best roomer I ever had; keeps his room spotless. Elsie don't have to do anything but make the bed. And I appreciate that, but if you ask me, it's kinda unnatural for a man to be so tidy. Isn't that what you say? There's been something suspicious about him from the very first.

Mrs. Hergesheimer. I made a point of talking to him when I was shopping in Baumgarden's yesterday. My, he struts around that floor. You'd think he was president instead of a floorwalker.[1] I asked him where they kept the artificial flowers. I knew, but I just wanted to see if he'd recognize me. He smiled and made a lordly gesture with his hand, showin' me the way. You'd have thought he was the king of Persia, with all his fine manners.

Mrs. Crosby. He belongs to the Lions Club. Do you think he'd be a Communist and still belong to the Lions Club?

Mrs. Hergesheimer. You can't tell. Lots of them join clubs like that just as a cover-up. That schoolteacher I got — she's a Red and I know it. Brings home all kinds of books to read. Yes. Dangerous books. But she goes to church every Sunday morning, just as big as you please, just to pretend she's *not* a Red.

Mrs. Crosby. Forevermore!

Mrs. Hergesheimer. I think you've got every right to go into that closet, Mrs. Crosby.

Mrs. Crosby. Yes, I think so too. Well — well, you come with me!

Mrs. Hergesheimer. Oh, Mrs. Crosby, honey, I don't think it's right for me to do it. I'll stay down here and see that Elsie doesn't bother you.

Mrs. Crosby. I'm not going to do it if you don't come with me.

Mrs. Hergesheimer. Well . . . .

Mrs. Crosby. After all, you've been just as curious about this as I've been, and I think you owe it to me to come along.

Mrs. Hergesheimer. Well, if that's the way you feel about it, Mrs. Crosby, I'll come along. After all, it's not as though we were doing anything criminal.

Mrs. Crosby. Indeed it's not. Come on, then.

[*She starts toward the stairs, taking a final look toward the kitchen to make sure that* ELSIE *is occupied.*]

[1]**floorwalker:** department-store employee who supervises sales personnel and assists customers.

Mrs. Hergesheimer [*following with some trepidation*]. Oh dear, I hope he doesn't find out.

Mrs. Crosby. We can get that lock off this time without making any more scratches than we made yesterday. He won't notice.

Mrs. Hergesheimer. Oh, I bet he does. He's got a sharp eye.

Mrs. Crosby. Well, I don't care if he does. I've got a right to see what's in that closet.

Mrs. Hergesheimer. Yes. Well — go on, Mrs. Crosby. I'm right behind you.

[*Slowly, cautiously, the two women go up the stairs together. The stage is empty for a few moments, then* ELSIE *comes in from the kitchen, looks up the stairs with curiosity. Then, as though the behavior of the two women were too much for her to understand, she shrugs her shoulders, laughs gently, and returns to the kitchen. The stage is empty again for a few moments. Then, slowly, the front door opens and* MR. NEWBOLD *returns inside the house. He has suspected the two women would do exactly what they're doing. He is very nervous. His heart is pounding. He starts up the stairs and then comes down again. He can't seem to get the courage to confront the women. The starch he showed earlier in the play has dissolved. He is perspiring heavily and twisting his hands in fear and excitement. In a few moments we hear the women on their way downstairs.* MR. NEWBOLD *hurriedly finds a closet to hide in. The women come down the stairs slowly, holding an awed silence.* MRS. CROSBY *carries a woman's hat, a large hat, the kind a graceful lady might wear to a garden party. It is quite a lovely hat, in a light pastel color with great flowers in its limber brim, and sleek satin ribbons. They come together,* MRS. CROSBY *holding the hat, both of them studying it with bafflement.*]

Mrs. Crosby. Hats! Dozens of hats! I can't believe it.

Mrs. Hergesheimer. He must have brought them home from the store, don't you think?

Mrs. Crosby. But there was all that sewing equipment on a shelf.

Mrs. Hergesheimer. But no man could make hats as lovely as these.

Mrs. Crosby. I don't know. There's something kinda unusual about Mr. Newbold. I think he might have made them. I — I know he did. [*She suddenly recalls a clue.*] I remember now how he was always looking through the fashion magazines. Sometimes he'd take them up to his room. He'd *study* them. I always wondered why.

Mrs. Hergesheimer. But why would he stay up in his room making hats — and then keep them locked in his closet?

Mrs. Crosby. He — he's just peculiar. That's all. He's just peculiar. I thought so, the first time I saw the man. He's too prim for a man. He's too tidy, the way he keeps his room. It's just not natural.

Mrs. Hergesheimer. Oh, I wish now we hadn't looked.

Mrs. Crosby. I had a perfect right.

Mrs. Hergesheimer. I know, but . . . .

Mrs. Crosby. Why, I think he's the most peculiar man I ever heard of. Why, I'd rather be harboring a Communist.

Mrs. Hergesheimer. Oh, Mrs. Crosby, don't say that.

Mrs. Crosby. I would. I'd rather be harboring a Communist than a man who makes hats.

Mrs. Hergesheimer. Why, there's nothing wrong with making hats. I don't see anything wrong with it. Why, lots of men make hats. Some of the finest designers there are are men. Why, of course.

Mrs. Crosby. But he kept them locked in his closet. He was ashamed of them.

Mrs. Hergesheimer. Maybe it's just a hobby with him. Some men knit, you know, because it helps their nerves.

Mrs. Crosby. I'm going to ask him to leave.

Mrs. Hergesheimer. Oh no, Mrs. Crosby. Don't do that.

Mrs. Crosby. I am. I'm going to ask him to leave. And I'm going to call the store he works at and tell them what kind of a freak they have working for them. Indeed I am.

Mrs. Hergesheimer. Oh, I wouldn't do that. It's not against the law for a man to make hats. He hasn't done anything really wrong.

Mrs. Crosby. Why, a man who'd make hats and lock them up in his closet, there's no telling what kind of a person he is. He might do any kind of dangerous, crazy thing.

Mrs. Hergesheimer. Oh, I don't think so, Mrs. Crosby. Really I don't.

Mrs. Crosby. I'd rather he was a Communist. At least you know what a Communist is up to. But a man that makes hats! What can you tell about such a creature?

Mrs. Hergesheimer. I wouldn't give it another thought, if I was you.

Mrs. Crosby. Well, I guess it takes all kinds of people to make a world.

Mrs. Hergesheimer. Of course. That's the way to look at it.

Mrs. Crosby. Hats! Hats! Hats! With *flowers* on them!

Mrs. Hergesheimer. I must run along now.

Mrs. Crosby. Hats!

Mrs. Hergesheimer. Goodness, I hope he never finds out. . . .

Mrs. Crosby. I don't care if he does. Just let him try to scold me, in that superior way of his. [*Imitating* MR. NEWBOLD.] "Mrs. Crosby, someone's been tampering with the lock on my closet. I demand privacy, Mrs. Crosby. That's all I ask, is just one tiny closet to call my own. That's all I ask." Hmm. I'll have an answer for him. "What in God's name does a grown-up man like you mean by making hats, Mr. Newbold? Shame!" That's what I'll tell him. And he won't act so superior then.

[MRS. HERGESHEIMER *flutters out of the house as though wanting to avoid further involvement.* MRS. CROSBY *studies the hat again, taking it to the mirror to try it on. She deliberately burlesques its elegance and all that it signifies of feminine daintiness and beauty. Then she tosses it onto a chair and returns to the kitchen. Stealthily,* MR. NEWBOLD *comes from the closet in which he has been hiding. He is a shattered man. His stiff pride, his erect authority, are destroyed. He picks up the hat lovingly, holding it in the air to admire again its loveliness. He even weeps. Finally, his courage restored, he determines on a course of action. Twisting a great beaded hat-pin from the feathers and furbelows*[2] *of his creation, he walks boldly to the kitchen door and stands, holding the pin behind his back, and calling in a voice that is eerie with its dire purpose.*]

Mr. Newbold. Mrs. Crosby! Could you come here a moment, please? Could you come here a moment, Mrs. Crosby?

[2]**furbelows** (fėr'bə·lōz): ruffles; showy decorations.

The Rabbits Who Caused

Within the memory of the youngest child there was a family of rabbits who lived near a pack of wolves. The wolves announced that they did not like the way the rabbits were living. (The wolves were crazy about the way they themselves were living, because it was the only way to live.) One night several wolves were killed in an earthquake and this was blamed on the rabbits, for it is well known that rabbits pound on the ground with their hind legs and cause earthquakes. On another night one of the wolves was killed by a bolt of lightning and this was also blamed on the rabbits, for it is well known that lettuce-eaters cause lightning. The wolves threatened to civilize the rabbits if they didn't behave, and the rabbits decided to run away to a desert island. But the other animals, who lived at a great distance, shamed them, saying, "You must stay where you are and be brave. This is no world for escapists. If the wolves attack you, we will come to your aid, in all probability." So

All the Trouble JAMES THURBER

the rabbits continued to live near the wolves and one day there was a terrible flood which drowned a great many wolves. This was blamed on the rabbits, for it is well known that carrot-nibblers with long ears cause floods. The wolves descended on the rabbits, for their own good, and imprisoned them in a dark cave, for their own protection.

When nothing was heard about the rabbits for some weeks, the other animals demanded to know what had happened to them. The wolves replied that the rabbits had been eaten and since they had been eaten the affair was a purely internal matter. But the other animals warned that they might possibly unite against the wolves unless some reason was given for the destruction of the rabbits. So the wolves gave them one. "They were trying to escape," said the wolves, "and, as you know, this is no world for escapists."

Moral: Run, don't walk, to the nearest desert island.

Vocabulary

admonished (ad mon′ishd) warned in a scolding manner.
"'Hold on tighter,' he admonished his hand." *The Wishing Well,* p. 41.

aggressive (ə gres′iv) self-assertive; showing drive and energy.
" . . . the dog . . . was growing more aggressive toward his ankles." *Tom Edison's Shaggy Dog,* p. 71.

aghast (ə gast′) shocked by something horrible.
"The Frenchman . . . stood aghast as though he, by accident, had killed a human being." *Friendship,* p. 10.

anguish (ang′gwish) deep emotional suffering.
" . . . all this . . . occasioned me moral anguish which I could not understand." *My Elder Brother,* p. 16.

apparition (ap ə rish′ən) a ghostly figure.
" . . . he stared at the dog . . . as if seeing an apparition." *Friendship,* p. 10.

apprehension (ap ri hen′shən) suspicious fear.
" . . . the mother moved back and forth . . . in an agony of apprehension." *The Use of Force,* p. 63.

arrogantly (ar′ə gənt lē) proudly; with great self-importance.
" . . . he blurted out arrogantly, like a strong, full-grown man, 'Just a minute.'" *All the Years of Her Life,* p. 85.

benefactor (ben′ə fak tər) a person giving a gift or aid.
" . . . seeing him eagerly wag his tail to greet his benefactor. . . . " *Friendship,* p. 9.

brusquely (brusk′lē) abruptly; bluntly.
"Sam Carr usually said, 'Good night,' brusquely, without looking up." *All the Years of Her Life,* p. 84.

burlesques (bėr lesks′) imitates mockingly; ridicules.
"She deliberately burlesques its elegance. . . ." *The Tiny Closet,* p. 142.

caddy (kad′ē) a boy hired to carry golf clubs and find golf balls for players; often spelled *caddie.*
"Beezie went home that night after listening to the caddy and lay awake. . . . " *The Wishing Well,* p. 33.

candid (kan'did) sincere; honest.
"He was impulsive, candid, and fickle in his enthusiasms." *My Elder Brother,* p. 17.

canine (kā'nīn) having to do with dogs.
" . . . man has no power whatever to change . . . their canine nature." *Friendship,* p. 12.

caress (kə res') fondle or pat.
"Lately he had begun to caress the dog. . . . . " *Friendship,* p. 7.

consoled (kən sōld') comforted.
"The Frenchman consoled himself that the dog would come again next day." *Friendship,* p. 11.

contemptible (kən temp'tə bl) deserving of scorn; low.
" . . . the parents were contemptible to me." *The Use of Force,* p. 63.

cordially (kôr'jəl ē) heartily; sincerely; warmly.
"The man had a cordially sinister air." *Ashes for the Wind,* p. 126.

culprit (kul'prit) guilty one.
"The real culprit, though, was the baker. . . . " *Friendship,* p. 7.

curios (kyůr' ē ōz) objects valuable for their oddness or rareness.
"Then it would be a rage for curios with which to adorn his table. . . . " *My Elder Brother,* p. 17.

deficiency (di fish'ən sē) lack.
"Must be a deficiency in his diet . . . . " *Tom Edison's Shaggy Dog,* p. 71.

deftly (deft'lē) skillfully.
"Deftly, the Frenchman snatched that steak from the platter of the astonished diner. . . . " *Friendship,* p. 10.

desisted (di zist'id) stopped; ceased.
"Perhaps I should have desisted and come back in an hour or more." *The Use of Force,* p. 64.

desolate (des'ə lit) bleak.
" . . . he thought how [the grin] had died on the dog's face and changed into a desolate sadness. . . . " *Friendship,* p. 11.

dire (dīr) having dreadful consequences; sinister.
" . . . a voice that is eerie with its dire purpose." *The Tiny Closet,* p. 142.

disconcerted (dis kən sėr′tid) disturbed; lacking in composure.
"The man was visibly disconcerted." *Ashes for the Wind,* p. 127.

diverse (də vėrs′) different; varied.
"Carried away by the most diverse interests, he flung himself into them heart and soul." *My Elder Brother,* p. 17.

docile (dos′l) easily managed; obedient.
"And she went like a docile, silent child to the photograph studio. . . . " *Lost Sister,* p. 109.

dominant (dom′ə nənt) controlling; superior.
" . . . she was becoming dominant there in the dimly lit store." *All the Years of Her Life,* p. 88.

drollish (drōl′ish) amusing.
" . . . the dog sitting there greeting the owner with his drollish grin." *Friendship,* p. 6.

dwindled (dwin′dld) decreased; lessened.
" . . . the pain slowly dwindled." *The Wishing Well,* p. 42.

eerie (ir′ē) frightening; weird.
" . . . a voice that is eerie with its dire purpose." *The Tiny Closet,* p. 142.

ensuing (en sü′ing) following.
"In the ensuing struggle they grew more and more abject. . . . " *The Use of Force,* p. 63.

esteem (es tēm′) high opinion.
"A small token of esteem, sir, for an ancestor of yours. . . . " *Tom Edison's Shaggy Dog,* p. 77.

evaded (i vād′id) avoided; escaped.
" . . . she evaded the followers she knew would come." *Lost Sister,* p. 114.

exotic (eg zot′ik) foreign; strange.
"They watched them like exotic birds from a green jungle." *The Wishing Well,* p. 34.

genial (jēn′yəl) warmly pleasant.
" . . . Mr. Carr began to feel warm and genial himself." *All the Years of Her Life,* p. 88.

gnarled (närld) knotted; twisted; rugged.
"Her gnarled hands must have been skilled at squaws' work. . . . " *Lost Sister,* p. 110.

guffawed (gu fôd') burst into loud, coarse laughter.
"'If I *want* to call him that?'—Bullard guffawed. . . ." *Tom Edison's Shaggy Dog,* p. 73.

hamlets (ham'lits) clusters of houses; villages.
"They . . . took two streetcars . . . the second one, ten miles into the country, through the little hamlets. . . ." *The Wishing Well,* p. 33.

harboring (här'bər ing) giving shelter to; protecting.
"For all I know, I may be harboring a spy. . . ." *The Tiny Closet,* p. 137.

hygienic (hī jē en'ik) sanitary; healthful.
". . . it wouldn't be hygienic or mannerly for a waitress to take each roll into her hand. . . ." *Friendship,* p. 7.

immersed (i mėrsd') covered by water.
"He stayed immersed [in the tank] all day long." *Not to Go with the Others,* p. 124.

impeccably (im pek'ə blē) in a faultless manner.
"He is always impeccably dressed. . . ." *The Tiny Closet,* p. 133.

impulsive (im pul'siv) acting on sudden desire.
"He was impulsive, candid, and fickle in his enthusiasms." *My Elder Brother,* p. 17.

indignation (in dig nā'shən) anger, of the type aroused by injustice.
"His face got red, and he knew he looked fierce with indignation." *All the Years of Her Life,* p. 84.

infinitely (in'fə nit lē) without limits; forever.
". . . the sadness of one who infinitely trusts somebody's honest friendship. . . ." *Friendship,* pp. 9–10.

inordinate (in ôr'də nit) much too great.
"But perhaps my inordinate sensitiveness and tendency to analyze deceived me in this case." *My Elder Brother,* p. 17.

insolence (in'sə ləns) boldness; rudeness.
". . . I talked back to someone with spoiled, small-boy insolence." *Lost Sister,* p. 108.

involuntary (in vol'ən ter ē) uncontrollable.
"An involuntary shiver straightened his whole body. . . ." *Friendship,* p. 10.

ledger (lej′ər) an account book.
"And I'll keep all guesses in a ledger till Christmas Eve. . . . " *Jug of Silver,* p. 46.

martyr (mär′tər) a person put to death or made to suffer greatly for his beliefs.
"Sabina . . . went in like a martyr, firmly clutching her Bible, and read aloud from it. . . . " *Lost Sister,* pp. 110–111.

matrons (mā′trənz) married women.
" . . . the matrons of the town kept their daughters under lock and key. . . . " *Jug of Silver,* p. 45.

molest (mə lest′) disturb or annoy.
"He stepped back a little from the door, as if not to molest patrons coming in or going out." *Friendship,* pp. 5–6.

patrons (pā′trənz) customers.
"The Frenchman [was] more interested in the dog than in his patrons . . . . " *Friendship,* p. 4.

perceptible (pər sep′tə bl) noticeable; visible.
"Who has not noticed those mysterious unspoken relations which manifest themselves in a barely perceptible smile . . . . " *My Elder Brother,* p. 16.

peremptorily (pər emp′tə rə lē) in a commanding or urgent manner.
"Bessie said something, then said it again, peremptorily . . . . " *Lost Sister,* p. 108.

premises (prem′is əz) property.
"He wanted him to . . . guard the premises night and day . . . . " *Friendship,* p. 7.

profusion (prə fū′zhən) large quantity.
"She had magnificent blonde hair, in profusion." *The Use of Force,* p. 61.

prostrate (pros′trāt) lying flat, face downward.
"SS men with flashlights waded among the bodies, shining lights in the faces of the prostrate victims." *Not to Go with the Others,* p. 121.

ravenous (rav′ən əs) extremely hungry.
"The bear was hungry, the bear was ravenous . . . . " *Adventures of Isabel,* p. 66.

serge (sėrj) a kind of cloth having slanting lines or ridges.
" . . . a new arrival in Florida, still buttoned up tight in heavy serge, stiff collar, and necktie . . . . " *Tom Edison's Shaggy Dog,* p. 71.

stealth (stelth) secrecy or slyness.
" . . . the Frenchman busied himself . . . all the while watching the dog by stealth . . . . " *Friendship,* p. 6.

stolid (stol'id) not easily excited; showing no emotion.
"Aunt Margaret recovered herself, and instead of embracing this silent, stolid woman, satisfied herself by patting an arm . . . . " *Lost Sister,* p. 103.

subsided (səb sīd'id) lessened in intensity; decreased.
"The ache subsided behind his eyes . . . . " *The Wishing Well,* p. 42.

sullen (sul'ən) gloomy; somber.
"The valley throbbed intact under the sullen morning." *Ashes for the Wind,* p. 127.

symmetry (sim'ə trē) regular, balanced arrangement.
" . . . I had upset the symmetry of the different treasures on his table." *My Elder Brother,* p. 17.

tremor (trem'ər) shaking; vibration.
"And without a tremor he most bravely took the shot." *Friendship,* p. 9.

trepidation (trep ə dā'shən) dread; fear.
Mrs. Hergesheimer follows "with some trepidation" in *The Tiny Closet,* p. 140.

turrets (tėr'its) small towers, sometimes found on the corners of buildings.
"Those who jumped or climbed down were shot . . . by SS men in the turrets on the walls . . . . " *Not to Go with the Others,* p. 122.

vermin (vėr'mən) small animals or insects, such as rats, fleas, and lice, that are troublesome or destructive.
" . . . her hair had been cut short to clean out the vermin." *Lost Sister,* p. 103.

voluble (vol'yə bl) extremely talkative; with a rapid flow of words.
" . . . when they told her, with signs and voluble explanations, to change her dress, she did." *Lost Sister,* p. 106.

Acknowledgments for Photographs and Illustrations

The photographs and illustrations in this book were obtained from the following sources: pp. i, ii, vi-vii, photos by Mark Silber; p. 1, photo by Margot Niederland; p. 3 (top), courtesy of Freelance Photographers Guild, photo by Harry Hartmann; (bottom), photo by Mark Silber; (right), photo by Chapin Davis; p. 5, photo by Chapin Davis; p. 8, courtesy of Pix, Inc., photo by Ann Zane Shanks; pp. 14-15, photo by Frank Siteman; p. 18, photo by Bill Sumner; pp. 20-21, 22-23, 25, 27, illustrations by Alfred Olschewski; pp. 28-29, courtesy of Magnum Photos, Inc., photo by Bruce Davidson; pp. 30-31, photo by Jay Bruner; pp. 32, 34, 38-39, 41, 43, illustrations by Thomas Upshur; pp. 44-45, 48, 50, 56-57, courtesy of Craven, Evans, & Stone, illustrations by Richard Taddei; pp. 60-61, courtesy of Design Photographers International, photo by Chris Reeberg; p. 67, photo by Olive Pierce; p. 68, photo by Christopher S. Johnson; pp. 70 and 72, courtesy of Nancy Palmer Photo Agency, photos by Michael Gold; p. 75, courtesy of Photoworld; p. 77, courtesy of Nancy Palmer Photo Agency, photo by Michael Gold; p. 78, photo by Chapin Davis; p. 81, photo by John Urban; p. 82, photo by Ernst Haas; p. 87, photo by George W. Gardner; p. 89, photo by Mark Silber; p. 91, courtesy of the Office of Economic Opportunity, photo by Paul Conklin; p. 95, photo by Paul Levasseur; p. 97, courtesy of Pix, Inc., photo by Paul Conklin; p. 99, photo by Benno Friedman; p. 100, courtesy of The National Archives, photo by Alexander Gardner; p. 104, photo by Paul Conklin; p. 106, photo by Paul Levasseur; pp. 107 and 111, photos by Paul Conklin; p. 112, courtesy of The Bureau of American Ethnology, Smithsonian Institute; p. 117, courtesy of Western Ways Photos, photo by Charles W. Herbert; p. 118, photo by Anna K. Moon; p. 122 (black and white), courtesy of Photoworld; (color), courtesy of Stock, Boston; p. 123, photo by Owen Franken; p. 125, photo by Bill Sumner; p. 126, courtesy of Louis Mercier, photo by Robert Dolman; p. 128, photo by Richard Bellak; p. 132, photo by John Posey; p. 135 (left), courtesy of Monkmeyer Press Photo Service, photo by Hugh Rogers; (right), photo by Eric Aerts; p. 139, photo by Eric R. Myrvaagnes; p. 143, courtesy of Stock, Boston. Cover, photo by Hans P. Widmer.

ABCDEFGHIJ–RM–798765421